Microbiological Assay for Pharmaceutical Analysis

A Rational Approach

Microbiological Assay for Pharmaceutical Analysis

A Rational Approach

William Hewitt

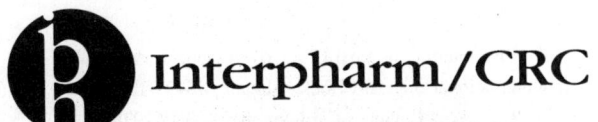
Interpharm/CRC

Boca Raton   London   New York   Washington, D.C.

FIRST INDIAN REPRINT, 2009

Library of Congress Cataloging-in-Publication Data

Hewitt, William, 1923–
  Microbiological assay for pharmaceutical analysis: a rational approach/William Hewitt.
    p. cm.
  Includes bibliographical references and index.
  ISBN 0-8493-1824-6
  1. Microbiological assay. 2. Microbiological assay--Mathematics. I. Title.
  [DNLM: 1. Biological assay--methods. 2. Data Interpretation, Statistical. 3. Dosage
  Forms--standards. 4. Drug Stability. 5. Microbiological Techniques. 6. Pharmaceutical
  Preparations--analysis. QV 771 H611ma 2003]
  QR69.M48H478 2003
  615′.19–dc22

2003061978

© 2004 by CRC Press LLC

Interpharm is an imprint of CRC Press LLC

No claim to original U.S. Government works
International Standard Book Number 0-8493-1824-6
Library of Congress Card Number 2003061978
Printed and Bound in India by Saurabh Printers Pvt Ltd.
FOR SALE IN SOUTH ASIA ONLY

Preface

One must have reasons for writing a book — the preface provides an opportunity to explain those reasons.

My background is that of an analytical chemist. By chance, I became involved in microbiological assay some decades ago and also became intrigued by the seemingly unusual calculations that were used to arrive at estimated potency and confidence limits for the estimate. I was fortunate in having excellent tutors — in particular, Peter Tootill, who had been prominent in promoting the use of good experimental designs in the 1950s and 1960s.

Some years later I embarked on an overseas career as an advisor on analytical chemistry in Europe. Inevitably my attention was drawn to the lack of understanding of the basic mathematical principles of microbiological assay (the agar diffusion assay for antibiotics). I set out to bring enlightenment in this field. This was first by some printed notes for local use. Later, recognizing the wider potential application of these notes, I was able to expand on them very substantially and published *Microbiological Assay: An Introduction to Quantitative Principles and Evaluation* in 1977 (book No. 1). It is fascinating to contemplate that this book, with its innumerable calculations, was assembled with the aid of a mechanical typewriter, a lot of correction fluid, and a pocket calculator.

Some years later, still working as an overseas advisor (and still very much an analytical chemist), I became even more impressed by the lack of appreciation of the basic practical awareness needed to achieve success in the field of microbiological assay. Managers were sometimes microbiologists — much more competent than myself in microbiology but not sufficiently aware that the microbiological assay of antibiotics needed to be treated as a branch of analytical chemistry. I recall one laboratory in which I found no fault with the microbiological aspect of the work, but in which the optical components of the zone measuring device were coated with the fine sands of a local desert. This and other similar observations prompted me to think about writing a book concerned with the practical aspects of microbiological assay. For this I needed a co-author; I was very fortunate in meeting Stephen Vincent of Glaxo Laboratories, U.K. Stephen had vast experience of running a microbiological assay laboratory; he agreed to collaborate with me. The outcome was *Theory and Application of Microbiological Assay*, Hewitt and Vincent, 1989 (book No. 2).

Book No. 1 appears to have served its purpose. In particular, it provided guidance on the mathematical aspects of microbiological assays in a form more readable than the pharmacopoeias. While the fundamental mathematical truths expounded in that book are as valid in 2003 as they were in 1977, the book is much outdated. It was written before the advent of personal computers. Because of the intense labor involved in a manual calculation of a statistical evaluation (typically one hour's work for an experienced person), such calculations were regarded as something not to be applied to each assay but as a demonstration of the precision that could be expected of an assay.

Nowadays, once the program has been set up on the computer, the calculation is effortless. This might seem to imply that the analyst need not concern himself or herself with the calculation, being satisfied that it has been set up by a competent statistician and validated. However, European pharmacopoeia guidance is otherwise. Referring to computer programs, the pharmacopoeia directs that the final decision to accept, reject, or modify the observed data must be taken by the analyst in the light of the detailed results given by the computer. In reaching a decision, the analyst should bear in mind that "statistical significance" is not synonymous with "practical importance" — a fact that is well illustrated in this book.

While the analyst will probably accept the principles of the statistical evaluation without having a thorough understanding of statistical theory, there is clearly a need for the analyst to understand how the calculations work. In the analysis of variance it is necessary to be able to identify why variation attributable to any source is statistically significant — then to assess whether that significance is really going to jeopardize the reliability of the assay. A major objective of this book is to provide an insight into how the calculations work and to help analysts make judgments on validity.

On the practical side, the objectives of the book are more limited. There exists excellent detailed guidance on practical aspects of the subject in Stephen Vincent's contribution to book 2. This is not outdated. The guidance given here draws attention to the microbiological aspects, which may not be so obvious to the chemical analyst, and to analytical aspects that may not be so obvious to the microbiologist. An important feature concerns the need for sterile equipment and aseptic techniques.

Few books have been written presenting an overall view of microbiological assay. The very comprehensive *Analytical Microbiology* edited by Frederick Kavanagh — volume 1 in 1963 and volume 2 in 1972 — remains of great value to those practicing microbiological assay today. It is hoped that *Microbiological Assay for Pharmaceutical Analysis: A Rational Approach* will serve its intended purpose in providing an update on earlier publications.

Author Biography

William Hewitt studied pharmacy at the University of Nottingham. Later he studied chemistry at the University of London. He worked in the pharmaceutical industry for several years as a quality control analyst; this included the antibiotic sector, where he was introduced to and became fascinated by the microbiological potency testing of antibiotics. This was followed by work overseas as an advisor in quality control of pharmaceuticals.

It was overseas that William recognized that the quantitative/mathematical principles of microbiological assay were not well understood. He prepared notes for use in his own laboratory explaining these principles. These were then expanded substantially and published in 1977 by Academic Press as: *Microbiological Assay: An Introduction to Quantitative Principles and Evaluation.*

Few books have been written on the subject of microbiological assay, and the need was perceived for a book with more emphasis on the practical and microbiological aspects of the subject. A second book was produced in 1989, written in collaboration with Stephen Vincent, whose meticulous approach to practical detail and staff training made him an ideal partner. *Theory and Application of Microbiological Assay* was published by Academic Press.

William has been active in promoting the understanding of microbiological assay through technology transfer in several Asian countries. He has also directed several short training courses in the U.K. and the U.S. and has acted as consultant to laboratories in Australia, Belgium, Ireland, Slovenia, and the U.K.

Now retired from active laboratory work, William continues to write and produce software for assay evaluation, *Hewitt Bioassays*. He lives in a delightful rural part of West Wales with his wife and three cats. He has a passion for traditional jazz and plays clarinet.

Acknowledgments

I should like to express my appreciation to the many organizations and people whose help and encouragement contributed to making this work possible.

Thanks are due to:

Academic Press for kindly assigning to the authors copyrights of two books, *Microbiological Assay: An Introduction to Quantitative Principles and Evaluation* (Hewitt) and *Theory and Application of Microbiological Assay* (Hewitt and Vincent).

The American Society of Microbiology for granting permission to abstract from tables in a publication by Fujiahara et al.

Boots Contract Manufacturing (U.K.) for granting permission to quote experimental data from their laboratories.

The British Standards Institution for granting permission to use extracts from BS1583 (1986) and BS1792 (1982) in Chapter 6. Complete standards can be obtained from BSI Customer Services, tel +44 (0) 208 996 9001.

Faulding Pharmaceuticals plc for granting permission to make reference to a collaborative assay of 1996.

The Journal of Pharmaceutical Sciences — Chapter 4 includes adaptations from the work of Garrett and Miller (1965). These adaptations are made with the kind permission of Wiley-Liss, Inc., a subsidiary of John Wiley & Sons, Inc.

The United States Pharmacopoeial Convention — Chapter 8 makes reference to USP tests for outliers, and Chapter 12 introduces the chi-squared test. The corresponding tables of annexes 3 and 9 are reprinted with permission. The United States Pharmacopoeial Convention, Inc. ©1999. All rights reserved.

The European Pharmacopoeia Commission — Extracts from the chapters on statistical analysis of four editions of the EP are quoted or reproduced with the kind permission of the European Pharmacopoeia Commission. In Chapter 6 there is one quotation; in Chapter 8 through Chapter 10 there are four quotations. Edition and page references are given in the reference list at the end of each chapter.

The World Health Organization — In Chapter 7 reference is made to the WHO publication "Accelerated Stability Studies of Widely Used Pharmaceutical Substances under Simulated Tropical Conditions" (1986). Table 7.1 and Table 7.2 are based on information abstracted from this publication with the kind permission of the World Health Organization. In Chapter 11 reference is made to a publication in the *Journal of Biological Standardization*. Annex 8 is taken from that publication and is reproduced verbatim with the kind permission of the World Health Organization.

The Society for General Microbiology for granting permission to reproduce figures from articles by Cooper et al. published in the *Journal of General Microbiology*.

I should like to thank the following individuals:

Louise Boddill, Susan King, and Tracy Whitchurch, of Boots Contract Manufacturing, U.K. for permission to use their observations on the turbidimetric assay of tyrothricin.

Andrew Broadbridge, formerly of the National Institute for Biological Standardization and Control, U.K. for information on the theory of zone formation and also on zone reading by image analysis.

Roger Dabbah of the United States Pharmacopoeial Convention for information concerning the legal status of the USP and the FDA in the U.S.

Edwin Hewitt, my late father (a very numerate nonscientist), for reviewing earlier manuscripts and making valuable suggestions on presentation.

Alan Holt of Southmead Hospital, Bristol, U.K. for carrying out the peer review and making valuable suggestions.

Roddy Morrison of the Library of the Royal Pharmaceutical Society of Great Britain for providing detailed answers to several questions.

Hans Nelis of the University of Ghent, Belgium for permission to quote his observations on the assay of polymyxin.

Gerald Shockman of the Institute for Cancer Research, Philadelphia, Pennsylvania, U.S. for permission to use his experimental data concerning the growth of organisms limited by the quantity of one essential amino acid and for allowing me to put my own interpretation on his findings.

Stephen Vincent, formerly of Glaxo Laboratories, U.K. for reviewing Chapter 6 and making valuable suggestions.

Table of Contents

1 Microbiological Assay in Perspective

BIOLOGICAL ASSAYS IN GENERAL

There are many types of biological tests, but for our purpose, a biological assay is defined as "a practical procedure whereby the potency of a material of unknown potency (the unknown) is estimated by comparison of its effect in a biological system with that of a reference standard of known or defined potency."

Although the subject of this book is *micro*biological assay (in which the biological system is a culture of a microorganism), it is important to compare and contrast it with *macro*biological assay (in which the biological system is a number of higher organisms, such as birds or mammals, or perhaps an isolated organ from such an animal).

Ideally, in any biological assay, it should be possible to consider the unknown to be a dilution of the reference standard in an inert matrix; also, ideally, the biological system should be homogeneous so that the unknown and reference standard would be reacting in identical situations. If these ideals could be attained, biological assay would be much simpler than it is in practice.

A biological system may consist of, for example, 24 small animals that would be divided into 4 groups of 6 animals. Each group would receive a different treatment: standard high dose, standard low dose, unknown high dose, and unknown low dose. Observed response might be, for example, the increase in weight of the animal after a defined number of days. Calculation of results would be on the basis of mean of observed responses of the six animals in a group.

Differences in the responses of individual animals of the four groups would arise for two reasons:

1. The different treatments lead to intended differences in response (this is the basis of the assay).
2. Innate differences in the six individual animals in a treatment group result in differing responses to that same treatment. This is known as *biological error.*

The effect of biological error is reduced in the following ways:

Clearly, the use of six animals per treatment rather than only one and calculation of estimated potency on the basis of mean responses to the four

1

    treatments are intended to reduce error by averaging out individual variation
    within the group.

    Animals are selected that are similar in terms of weight, sex, and strain.

    Allocation of animals to a group would be on a completely random basis,
    such as by numbering the animals and using a table of random numbers to
    select animals for each group.

    The effect of biological variation may be further minimized by the use of
    crossover tests in which the test is repeated but the two groups that received
    the standard doses in the first test receive the unknown doses in the second
    test.

Despite such precautions, biological error remains. A measure of the extent of
such variation is obtained through application of elaborate mathematical techniques
in the statistical evaluation. That application leads to the estimation of the sample's
potency together with confidence limits showing the probable range within which
the true potency may lie.

Such statistical techniques (although not the crossover test) are described in
pharmacopoeias for processing of the raw data obtained in *micro*biological assays,
and, perhaps not surprisingly, the notion of biological error has stuck.

The number of living organisms introduced into a single test unit in a microbiological assay may be measured in tens of thousands or tens of millions, depending
on the particular type of assay. Furthermore, these individuals are probably fairly
uniform in their reaction to the active substance, and so it seems unlikely that
biological error is a big feature in microbiological assays.

There are certainly sources of error, but it seems unlikely that they have much
biological component. The nature of errors in microbiological assay will be discussed
in later chapters.

THE DIFFERENT SORTS OF MICROBIOLOGICAL ASSAY

The main use of microbiological assay is in the determination of the potency of
growth-inhibiting substances — especially antibiotics — and of growth-promoting
substances — amino acids and vitamins of the B group. There are three main assay
procedures:

1. The agar diffusion assay, which is applicable to both growth-inhibiting
   substances and growth-promoting substances
2. The flask or tube assay for growth-promoting substances
3. The tube assay for growth-inhibiting substances

Although methods 2 and 3 may appear similar, the bases of the assays and practical
techniques are rather different. These three methods are outlined in the sections that
follow.

THE AGAR DIFFUSION ASSAY

This assay method was devised by Heatley, who described it in a publication in 1944. It had been in use in his laboratory at the Sir William Dunn School of Pathology, Oxford since 1940. The ideal dose-response relationship for the agar diffusion assay for both antibiotics and vitamins is shown in Figure 1.1. A plot of the response (mean zone diameter) against logarithm of dose gives a line that is straight or very nearly straight. The true nature of this dose-response line will be discussed in greater detail in Chapter 2, where it will be seen that in theory and in practice the line is not straight. For the present, the approximation is good enough for most assay purposes.

Figure 1.1 represents an assay in which there are two preparations, the reference standard and an unknown sample. This is a symmetrical assay, i.e., each preparation has the same number of dose levels (three) in the same ratio to one another (2:1).

As the potencies of samples are unknown before assay, their responses must be plotted against a nominal dose. Each sample test solution is allocated a nominal dose identical with that of the corresponding standard test solution. When mean observed responses to the sample test solutions are plotted against the logarithms of their nominal potencies, lines that are parallel to the standard log dose-response line result. For sample test solutions more potent than their nominal potencies, the line lies above that of the reference standard. Conversely, when less potent, the line lies below that of the standard. Such assays are known as *parallel-line assays*. The relative potency of unknown to standard is obtained effectively from the horizontal distance between the two lines. In this case the horizontal distance in Figure 1.1 is 0.0547 on the log scale, so that the potency of the unknown relative to that of the

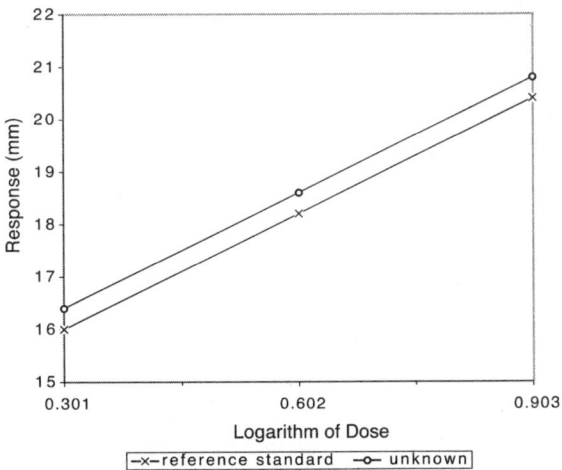

FIGURE 1.1 Ideal log dose-response lines for a three-dose level parallel-line assay. The line for the unknown is above that for the reference standard, thus indicating that the unknown test solutions are more potent than those of the reference standard.

standard is given by antilog 0.0547 = 1.13. Thus, the unknown has a potency of 113.4% of the standard. (Normally the concentration of the unknown test solution is adjusted so that relative potency is closer to 100%).

THE TUBE OR FLASK ASSAY FOR GROWTH-PROMOTING SUBSTANCES

An assay for riboflavine using lactic acid bacteria was described by Snell and Strong (1939). The principle of the assay was that the nutrient medium contained an abundance of all nutrients essential for the growth of the organism except riboflavine, which was absent. On adding small amounts of riboflavine to tubes, growth would take place on incubation, with the extent of growth dependent on the quantity of riboflavine added.

An almost ideal dose-response pattern for a growth-promoting substance assay by the tube method is shown in Figure 1.2. In this case, responses to standard test solutions are plotted against dose itself. Ideally, the dose-response line is a straight line passing through a point corresponding to zero dose and zero response.

As in the case of the agar diffusion assay, because sample potencies are unknown, test solutions are allocated nominal potencies equal to those of the corresponding standard test solutions. Plotting observed responses against nominal potencies leads to lines that ideally are straight and also pass through the point corresponding to zero dose and zero response. However, as illustrated in Figure 1.2, the response to zero dose is not always zero; it may have a small positive or negative value. In this case, a small positive value of two arbitrary units is shown. This might be due to a small amount of the active growth-promoting substance present in the nutrient medium.

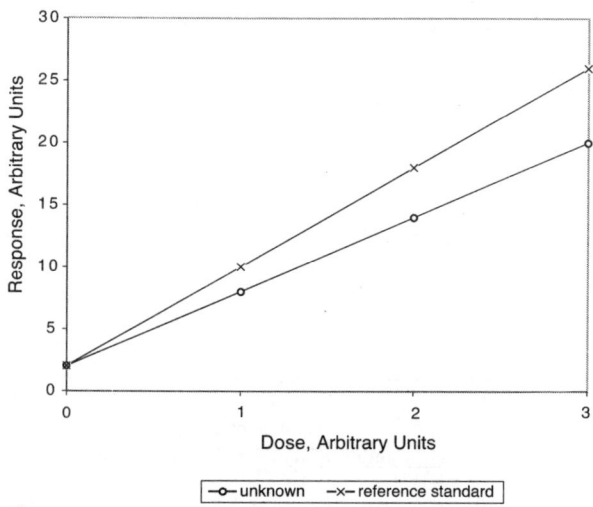

FIGURE 1.2 Ideal log dose-response lines for a three-dose level slope ratio assay. The line for the unknown is below that for the reference standard, thus indicating that the unknown test solutions are less potent than those of the reference standard.

As shown in Figure 1.2, for sample test solutions more potent than their nominal potencies, the line lies above that of the reference standard. Conversely, when less potent, the line lies below that of the standard.

The quantitative principle of the assay is that the ratio of the slopes of two lines is the same as the ratio of the potencies of the corresponding test solutions. Such assays are known as *slope ratio assays*. The relative potency of the unknown is obtained as the ratio *ca:ba* where *b* and *c* are the intercepts of a vertical line on the two response lines and *a* is the response to zero dose. Criteria for validity are that the two response lines are straight and intersect at zero dose.

Although Figure 1.2 shows the almost ideal case, in fact the lines are often curved. The reasons for and the effect of deviations from the ideal will be discussed in Chapter 3.

THE TUBE ASSAY FOR GROWTH-INHIBITING SUBSTANCES

The first published account of this assay system appears to be that of Foster (1942), who described an assay for penicillin. An idealized form of the dose-response line for a growth-inhibiting substance tube assay is shown in Figure 1.3. Here, response (turbidity) is plotted against logarithm of dose. At zero dose and very low doses of growth-inhibiting substances, growth of the organism is not inhibited. The resulting turbidity is taken as 100% growth. This is represented by the range *a* to *b* in Figure 1.3. At the other extreme, sufficiently high doses of growth-inhibiting substances will completely inhibit the growth, and so the turbidity in such tubes (range *e – f*)

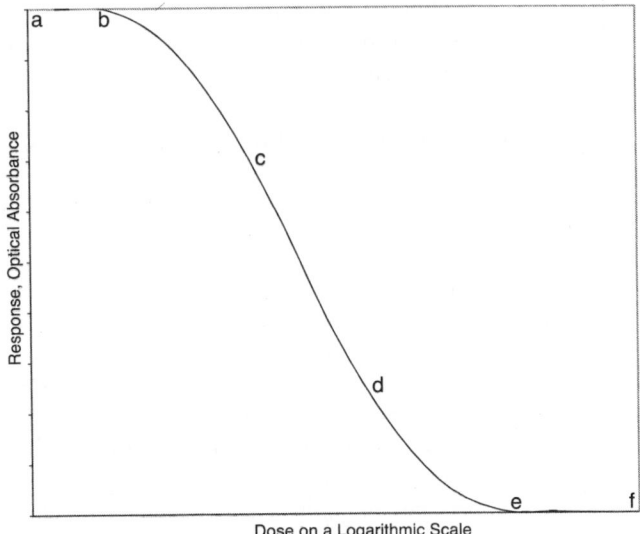

FIGURE 1.3 A representation of the form of log dose-response lines for a turbidimetric assay of a growth-inhibiting substance. This representation shows the entire curve for the standard preparation only. In practical assays, only a part of the curve is seen because only the central part (*c* to *d*) is useful.

or in an uninoculated blank is taken as 0%. Intermediate doses of growth-inhibiting substances lead to varying degrees of inhibition, so that a plot of response against log dose is sigmoid.

Once again, because the sample was allocated a nominal potency (which will not generally coincide with its true potency), a plot of response against nominal log dose will produce a curve displaced to the left or right of that for the standard for samples that are respectively more potent and less potent than their nominal potencies.

Sigmoid curves do not provide a very convenient basis for arithmetical derivation of the potency estimate. In practice, the entire curve is never seen, because doses are chosen to obtain responses corresponding to the central part of the line represented by c to d.

By choosing a part of that central portion that is straight (or nearly straight), the assay can be treated as a parallel-line assay, and results can be calculated as for the agar diffusion assay.

THE BASIS OF CALCULATION OF POTENCY ESTIMATES

Like the assays themselves, the calculations are based on ideals that may not correspond exactly with the actual situation. Basically there are only two models for the calculations:

1. The response (or a function of response) is directly proportional to the logarithm of the dose of active substance.
2. The response (or a function of response) is directly proportional to the dose itself of active substance.

These two models are the basis of parallel-line assays and slope ratio assays, respectively.

PARALLEL-LINE ASSAYS

Consider a simple assay having three dose levels each of both standard and unknown, in which the mean response (e.g., mean zone diameter for any individual treatment) is directly proportional to the logarithm of dose. If mean response to each reference standard dose is plotted against dose on a logarithmic scale, the log dose-response line will be straight. Because the potency of the unknown is in fact unknown, a true log dose-response line cannot be plotted. However, the mean responses can be plotted against the nominal potency, which is arbitrarily fixed at the same as that of the reference standard. This is illustrated in Figure 1.1, in which the line for the unknown is parallel to that of the standard and is above it, thus indicating a true potency greater than that of the reference standard. The quantitative potency relationship is derived essentially from the horizontal difference between the two lines on a logarithmic scale; the actual calculation procedure will be illustrated in Chapter 9.

SLOPE RATIO ASSAYS

Consider another simple assay having three-dose levels each of both standard and unknown, as well as a zero-dose control — seven treatments in all. The mean response for any individual treatment (e.g., turbidity due to microbial growth in a liquid medium) is directly proportional to the dose itself. If mean response is plotted against dose (on an arithmetic scale), the dose-response line will be a straight line, which should also pass through a point corresponding to zero dose and zero response, as illustrated in Figure 1.2.

Again, as the potency of the unknown is in fact unknown, its true dose-response line cannot be plotted. However, the mean responses can be plotted against the nominal potency, which is arbitrarily fixed at the same as that of the reference standard.

As seen in Figure 1.2, the line for the unknown is a straight line also passing through the zero-dose, zero-response point. The relationship between the potencies of the unknown and the reference standard is the same as the ratio of their slopes, hence the name slope ratio assay. In principle, any number of dose levels may be used, and it is not essential that there be the same number of levels for standard as for unknown. However, it will be shown later that it is better to use the same number of dose levels for each preparation (a symmetrical assay) and that there is no advantage in using a large number of levels. This particular assay, having three dose levels of each preparation, as well as the zero-dose control, may also be described as a seven-point common-zero assay. The calculation procedure will be illustrated in Chapter 10.

WHICH MATHEMATICAL MODEL?

The foregoing descriptions show the principles of the two mathematical models. In fact, the calculation procedures used are complex. They take into consideration random error and assess the extent of deviations from the model. Deviations that are too great might suggest invalidity of the assay.

The applicability of the two models is as follows:

1. Agar diffusion assays, whether of growth-promoting or growth-inhibiting substances, give responses that are treated as parallel-line assays. The lines may not be quite straight, but they should be parallel for a valid assay.
2. Tube assays for growth-promoting substances may be treated as slope ratio assays. However, for reasons that will be explained later, if curvature is significant, it will be necessary to abandon the slope ratio calculation and obtain a potency estimate graphically by interpolation from the standard curve.
3. Tube assays for growth-inhibiting substances are a more complex case. The full log dose-response relationship is represented here by a sigmoid curve as shown Figure 1.3. (A more exact description of the shape is

suggested in Chapter 4.) In practice, the full curve is never seen. A central portion of the curve may approximate sufficiently to a straight line that the parallel-line model is applicable. Mathematical transformations have been suggested to straighten curves. These are discussed in Chapter 4; however, it appears that these are little used. The general experience is that that slight curvature has little ill effect in *symmetrical parallel-line* assays.

REFERENCES

Foster, J.W. 1942. Quantative estimation of penicillin, *J. Biol. Chem.*, 144, 285–286.
Heatley, N.G. 1944. A method for the assay of penicillin. *Biochem. J.*, 38, 61.
Snell, E.E. and Strong, F.M. 1939. *Ind. Eng. Chem. Anal. Ed.*, 11, 346.

2 The Agar Diffusion Assay — Its Quantitative Basis

HISTORICAL INTRODUCTION

Penicillin was discovered by Alexander Fleming in 1929. It was in 1939 that a program of research began with a view to possibly manufacture penicillin or other substances having antibacterial activity. This project was at the Sir William Dunn School of Pathology in Oxford, England and was directed by Howard Florey and Ernst Chain. Penicillin proved to be the most promising candidate substance.

Norman George Heatley, a biochemist, was assigned the task of developing conditions of growth of the penicillin-producing fungus so as to obtain enough active substance for realistic investigation of its medical potential. There were already available some cumbersome, semiquantitative assay methods to monitor the processes of fermentation, extraction, and purification, but these were inadequate for the routine monitoring of many samples. In 1940, Heatley devised the agar diffusion assay for penicillin.

Heatley (1944) described the methods that had been developed and used in his Oxford laboratory since 1940 with only slight modifications:

Sterile nutrient agar was poured into a petri dish and allowed to cool and set. Plates were seeded by pouring on a broth culture of the test organism (*Staphylococcus aureus*). After tilting and shaking the plate to spread the culture it would be left at an angle of 20° to the horizontal to allow the cultures to drain for a short time then the surplus suspension was drawn off. The edge of the plate was marked at the point where the surplus liquid had been drained off. The plates were then dried in an incubator for 1–2 hours. Cylinders of glass or vitreous porcelain were sterilised by dry heat in a petri dish. They were picked from the dish with forceps, then momentarily flamed so that on placing on the agar surface a fluid-tight seal was produced by melting of the agar at the point of contact. Four cylinders were placed on each plate. Test solutions were added to fill the cylinders (although it had been observed that the exact volume seemed to make little difference). Plates were incubated overnight. The resultant inhibition zones were measured with pointed calipers then the distance between the two points of the caliper was read from a paper scale (which was burned after use) or a metal scale which could be flamed.

It was observed that when a single test solution was applied in different positions around the plate, zones from the edge from which the surplus broth culture had been drawn were sometimes smaller than those at the opposite edge. The difference in diameter was sometimes as much as 2 mm. The reason was not understood at the time.

It was because of these differences that the point of drainage was marked. Errors from this source were minimized in a quadruplicate test by placing the solutions in the same relative positions on each plate but rotating each successive plate through 90° before applying the test solutions. It had also been noted that the zone sizes and shape of the curve that related dose to zone size were not reproducible from day to day.

The need for a daily standard curve was apparent as was the need for a stable reference standard. The interesting history of reference standards is described in Chapter 7.

THE THEORY OF ZONE FORMATION

Inhibition zones are formed through the interaction between the growth-inhibiting substance diffusing through a nutrient agar gel and the increasing population of the sensitive organism with which the gel has been inoculated. To understand zone formation, it is necessary to visualize what is happening in physical, chemical, and biological terms.

The main constituent of agar is the calcium salt of a sulphuric ester of the polysaccharide, agarose. This consists of long chains of galactose units with α and β linkages. Nutrient agar medium consists typically of 1.5% agar and 1.5 to 2.0% of other dissolved substances, which include nutrients and perhaps salts. Thus, the medium has a water content of over 96%. The agar forms a continuous lattice imparting rigidity to the gel. The lattice and the dilute aqueous solution are inter-woven so the solution is a continuous phase through which solutes may diffuse almost as freely as through water, provided that the molecules are not large compared with the pore size of the lattice and also provided that no interaction exists between the solute and the agar lattice. The pore size of the lattice may be, for example, about 3 mμ, which is large compared with the size of a molecule of most antibiotics.

The scenario of zone formation is a solution of the growth-inhibiting substance, the test solution, being placed in a reservoir in contact with nutrient agar medium, which is inoculated uniformly with a test organism that is sensitive to the growth-inhibiting substance. In the plate assay, the agar medium is a layer of uniform thickness, perhaps 3 or 4 mm thick. The reservoir may be:

A cylindrical hole cut into the agar
A stainless steel cylinder placed on the surface of the agar, making a water-tight seal with the surface
A porcelain bead (electrical insulator fish-spine bead) placed on the surface of the agar after having been dipped into the test solution to take up liquid by capillary action
A circular disc of filter paper used in a similar manner to the porcelain bead

During incubation, the active substance (growth-inhibiting substance) diffuses into the agar medium; at the same time, in areas of the plate that the growth-inhibiting substance has not reached, the test organism population increases until it attains a level at which the growth-inhibiting substance is absorbed and so cannot advance further. At this time, known as the critical time, the position of the zone boundary is fixed. Further growth of the test organism leads to opacity of the medium around clear zones where growth has been inhibited.

The quantitative theory of microbial inhibition is discussed in Chapter 4 with respect to turbidimetric assays. The same principles apply to inhibition in an agar medium. Suffice it to say here that there is a range of concentrations in which growth will be reduced before the concentration that causes complete inhibition. It could be envisaged, therefore, that the zone boundary would be diffuse. In fact, zone boundaries are often quite sharp.

One possible explanation for sharp boundaries is that when the zone boundary has been established and vigorous growth takes place in the uninhibited areas, nutrients and oxygen will begin to be depleted. However, nutrients and oxygen will diffuse outwards from the inhibition zones, leading to increased growth at the boundary and, thus, to sharper zone boundaries.

Although agar is rather inert, there are possibilities of interaction with the growth-inhibiting substance. It may act as a weak ion exchanger so that diffusion of the growth-inhibiting substance could be made slower by a chromatographic effect. This would reduce the slope of the assay and cause problems if the growth-inhibiting substance were a mixture of active substances. The inclusion of salts in the medium would reduce the ion exchange activity.

The quantitative principles involved in the formation of inhibition zones in the antibiotic agar diffusion assay were studied by several workers during the years 1946 to 1952.

Much work was done using vertical narrow tubes (3 mm diameter) containing inoculated nutrient agar with antibiotic solution applied over the agar. This system had the advantage that temperature could be controlled precisely by placing the tubes in a water bath.

Mitchison and Spicer (1949) used narrow tubes of medium inoculated with staphylococci for the routine assay of streptomycin. The degree of anaerobiosis produced did not affect the minimum concentration of streptomycin required to inhibit growth. They concluded on both theoretical and practical grounds that the square of the width of the inhibition zone was directly proportional to the logarithm of the concentration of streptomycin.

This tube technique was adopted by Cooper and Gillespie (1952, 1) to study the influence of temperature on zone formation, and by Cooper and Linton (1952, 8) to compare the results obtained in tubes with those from plates. Zones in the two systems are illustrated in Figure 2.1. Work in this field is reviewed by Cooper (1963, 1972) in his contributions to Kavanagh's *Analytical Microbiology*, Volumes I and II. These studies consider the mathematics of diffusion of the antibiotic through the agar gel, growth pattern of the test organism, and interaction of the antibiotic with the living cell.

Cooper demonstrated certain mathematical concepts of importance for an understanding of the fundamental principles of this assay method. These concepts are very relevant to assay design and technique. These are some of the terms used in the equations that follow:

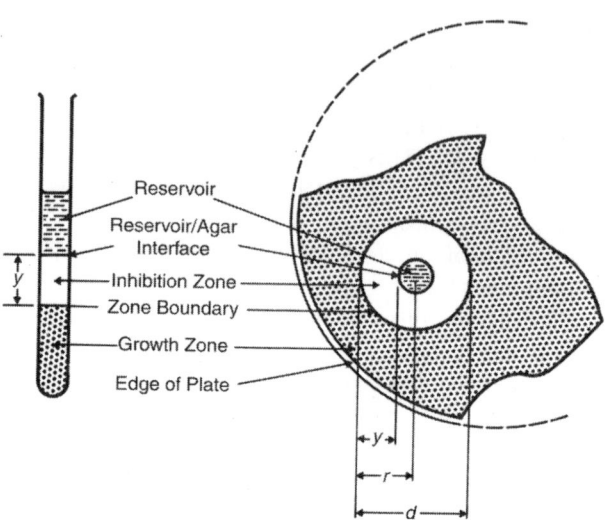

FIGURE 2.1 An illustration of inhibition zones formed in inoculated agar medium in tubes and in dishes:

y is zone width (tubes and dishes)

r is zone radius

d is zone diameter

Initial concentration m_0 — the initial concentration of antibiotic in the reservoir.

Critical concentration m' — the concentration of the antibiotic arriving at the position of the future zone boundary at a certain time T_0 (see below). (Note that in some original publications, the symbol σ is used for critical concentration.)

Zone width y — the distance between the edge of the reservoir and the zone boundary. (Note that in some original publications, the symbol x is used for zone width.)

Critical time T_0 — the period of growth of the organism at which it reaches the critical population N' (see below).

Inoculum population N_0 — the population at the time of inoculation.

Critical population N' — the population at time T_0 (further increase in population beyond this limit results in an excess of organism capable of completely absorbing the antibiotic and thus preventing its further outward diffusion; however, diffusion of the antibiotic during the lag phase of growth may result in small inhibition zones, even if a very heavy inoculum were used such that $N_0 = N'$).

Inhibitory population N'' — the population that is just sufficiently large to completely prevent formation of inhibition zones.

The nature of some of these parameters is now described in more detail.

Critical concentration m' is a measure of the sensitivity of the test organism under the particular assay conditions. It is not the same as minimum inhibitory concentration (MIC), being about two to four times as great as MIC, which is determined under very different conditions.

It is defined mathematically by

$$\ln m' = \ln m_0 - y^2/4DT_0 \tag{2.1}$$

where D is the diffusion coefficient (expressed as mm/hour), which is dependent on temperature and viscosity of the medium and varies inversely as the radius of the molecule.

Critical concentration may be evaluated by plotting y^2 against $\log m_0$ (see Figure 2.2). The intercept of the straight line on the $\log m_0$ scale at $y^2 = 0$ is 0.95 and corresponds to $\log m'$ so that m' is 8.9 µg/ml. It is clear that concentrations below m' cannot produce inhibition zones.

Critical time T_0 is the time at which the position of the zone boundary is fixed. It may be determined by preincubation of the inoculated medium prior to addition of the antibiotic solution to the reservoir. Critical time is defined mathematically by

$$T_0 = h + y^2/4D \ln(m_0/m') \tag{2.2}$$

where h is the time of preincubation.

Equation (2.2) is obtained by rearranging Equation (2.1) and replacing T_0 by ($T_0 - h$), which is a legitimate change because preincubation has effectively changed N_0.

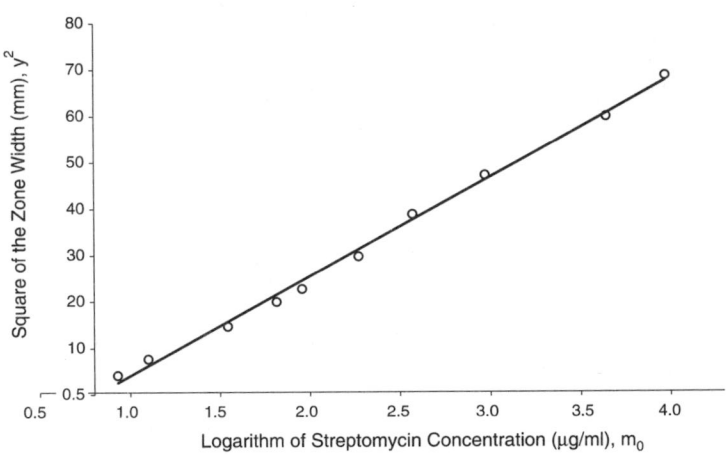

FIGURE 2.2 The relationship between square of width of inhibition zone and concentration of antibiotic. This is the work of Cooper and Gillespie (1952). It refers to inhibition zones produced by varying concentrations of streptomycin on nutrient agar medium in tubes inoculated with a strain of staphylococcus. The intercept of the line on the horizontal axis corresponding to zero zone width is at $\log m_0 = 0.95$ and corresponds to the critical concentration m'. Thus, $m' = $ antilog $0.95 = 8.9$ µg/ml.

Critical time T_0 may be evaluated by plotting y^2 against h for a fixed concentration m_0. The intercept of the straight line thus obtained on the h scale at $y^2 = 0$ corresponds to $h = T_0$.

Cooper (1963, 17) has shown that repetition at different concentrations *of* m_0, gives the same value for T_0 (see Figure 2.3).

Critical time is dependent on the lag period L and the generation time G. It is, therefore, temperature dependent. It is also dependent on the inoculum level, N_0. It is independent of the antibiotic concentration m_0.

To summarize, the factors ultimately deciding the position of the zone boundary are critical concentration and critical population. Factors that influence the diffusion of the antibiotic or the time of achieving the critical population, therefore, affect the zone size.

Cooper (1963, 51) reported a study using staphylococcus and streptomycin, using narrow tubes of inoculated nutrient agar to show the effect of differing inoculation levels on zone size. Eighteen different inoculum levels, N_0, ranged from 2.1×10^3 to 5.4×10^8 or expressed as $\log_2$, from 11 to 29; streptomycin concentrations m_0 were 10, 100, and 1000 µg/ml.

A plot (Figure 2.4) of square of zone depth, y^2, versus inoculum level gave three straight lines converging at y^2 when $N_0 = 1.34 \times 10^8$ or $\log_2 N_0 = 27$. This is the inhibitory population, N''.

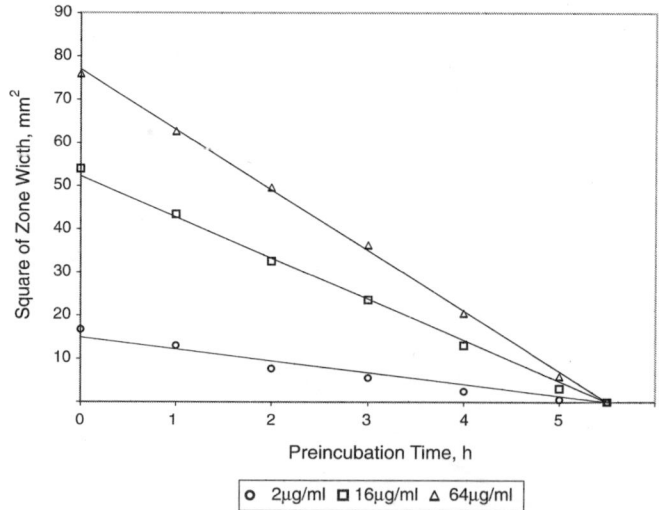

FIGURE 2.3 This is the work reported by Cooper (1963, 17). It shows the effect of incubation prior to addition of antibiotic solution to reservoirs. It refers to inhibition zones produced by varying concentrations of streptomycin on nutrient agar medium in tubes inoculated with a strain of staphylococcus. The three lines intersect the horizontal axis at the same point, 5.36 hours, which is the critical time T_0.

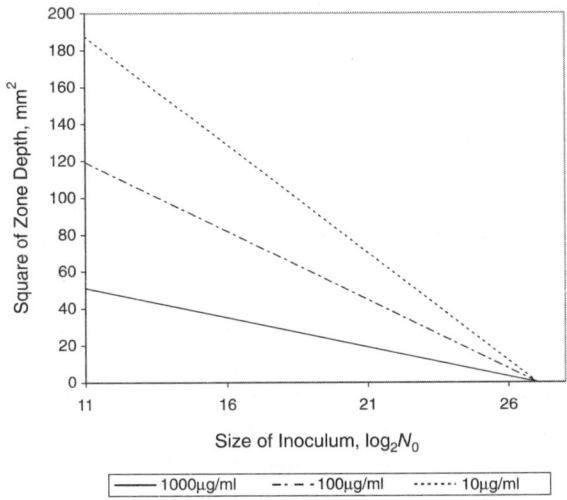

Size of Inoculum, $\log_2 N_0$

| ——— 1000µg/ml | – · – · 100µg/ml | ······· 10µg/ml |

FIGURE 2.4 This is the work reported by Cooper (1963, 51). It shows the effect of size of the inoculum on zone width for three concentrations of antibiotic. It refers to inhibition zones produced by streptomycin on nutrient agar medium in tubes inoculated with a strain of staphylococcus. The three lines intersect the horizontal axis corresponding to zero zone size at the same point, $\log_2 27$, corresponding to an inhibitory concentration N'' of about 1.34×10^8.

The general system considered by various workers comprised:

1. A reservoir from which the antibiotic was able to diffuse outward through the agar gel, thus inhibiting growth.
2. A nutrient agar medium inoculated with a uniform suspension of a test organism, either in the vegetative or spore form. This, on incubation, after a lag period — or in the case of spores, a germination plus lag period — multiplied to a level at which there was sufficient cell material to absorb all antibiotic, thus preventing further outward diffusion of antibiotic and so limiting the size of the inhibition zone.

Several mathematical models have been devised for both linear diffusion and radial diffusion from a reservoir. Two representative models are shown here. Those for linear diffusion apply in the case of diffusion in a tube and in the plate method when the reservoir is relatively large (8 mm diameter or more). They assume a constant concentration (m_0) of antibiotic; that is, the reservoir is sufficiently large that there is no appreciable drop in concentration of antibiotic during zone formation.

Those for radial diffusion are applicable in the plate method to small reservoirs, beads, or discs applied to the surface. They allow for falling concentration of antibiotic in the reservoir as diffusion proceeds.

For linear diffusion, Cooper and Woodman (1946) derived the equation

$$y^2 = 2.303 \times 4Dt(\log m^0 - \log m') \tag{2.3}$$

This showed that a plot of square of zone width against logarithm of antibiotic concentration would be rectilinear.

An equation derived by Mitchison and Spicer (1949) indicated the same relationship when the zone width was greater than 3 mm. However, at zone widths less than 3 mm, a plot of zone width unsquared against logarithm of antibiotic concentration was rectilinear.

For radial diffusion, Vesterdal (1947), considering small cups on the agar surface, and Humphrey and Lightbown (1952), considering small beads on the agar surface, derived slightly differing equations. Both showed that a plot of square of zone radius (r) against logarithm of antibiotic concentration was rectilinear.

Humphrey and Lightbown derived the equation

$$r^2 = 2.203 \times 4Dt(\log m_0 - \log 4\pi hDtm') \tag{2.4}$$

The last term in this expression includes h, the thickness of the agar, and shows that increasing the thickness of the agar will reduce zone diameter.

The other equations do not consider thickness of the agar, yet Lees and Tootill (1955), referring to wells punched in the agar, list thickness of the agar medium as one of the factors affecting zone size.

When the reservoir is small, the concentration in the reservoir soon drops below its original value (m_0). Instead of a continuously decreasing concentration at increasing distances, the reservoir is surrounded by an expanding concentric peak concentration. This may lead to double inhibition zones.

The position is somewhat confused by different workers using (in some instances) different symbols for the same parameter and by the fact that the diffusion coefficient D may be expressed as cm^2/sec or mm^2/h. The latter seems more appropriate. The relationship is exemplified by the data of Cooper and Woodman (1946) for the diffusion of crystal violet from a 0.2% aqueous solution into 2.5% nutrient agar at 15°C, for which the diffusion coefficient was

$$D = 3.02 \times 10^{-7} \, cm^2/sec = 3.02 \times 10^{-7} \times 60 \times 60 \times 100 = 0.109 \, mm^2/h$$

WHAT HAPPENS IN PRACTICE

Hewitt (1977) used the data of Cooper to see the effect of inoculum level on slope. Vertical lines were drawn on Cooper's graph (Figure 2.4) corresponding to $\log_2 N_0 = 11, 15, 19,$ and 23 to obtain the intercepts on the three lines representing 10, 100, and 1000 µg/ml. These intercepts showed the relationship between log dose and response for each of the four inoculum levels. These values are plotted in Figure 2.5, from which it is seen how a lower inoculum level gives a steeper slope.

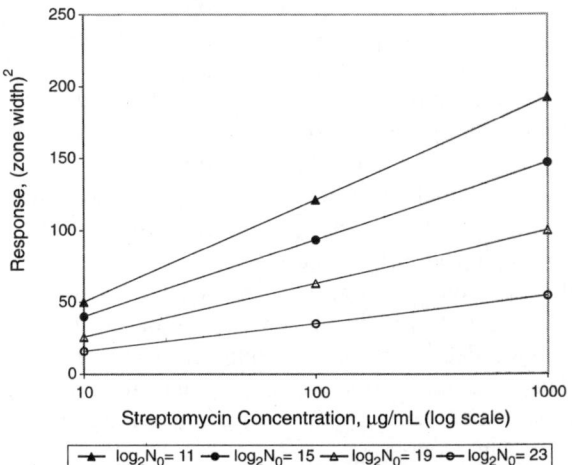

FIGURE 2.5 This graph is derived from the data of Figure 2.4. Vertical lines were drawn on that graph corresponding to $\log_2 N_0 = 11$, 15, 19, and 23 to obtain the intercepts (square of zone width) on the three lines representing 10, 100, and 1000 µg/ml. Square of zone width was then plotted against concentration of streptomycin for each inoculum level. This illustrates how slope of the log dose-response line varies with inoculum level.

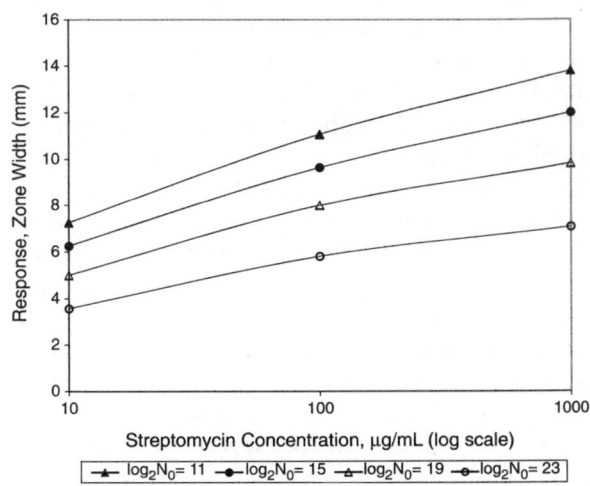

FIGURE 2.6 This graph is derived from the data of Figure 2.5. Zone width replaces square of zone width. This illustrates the curvature arising from using zone width unsquared. Similarly, a curved response line may be expected when zone diameter (unsquared) is plotted against logarithm of dose.

To illustrate what happens when we use zone diameters instead of square of zone width, the unsquared zone widths are plotted in Figure 2.6. Curvature of the response line is very apparent over the whole 100:1 range for all inoculum levels. However, it can be seen that curvature is not so apparent over the short ranges 2:1 and 4:1 that are commonly used in routine assays.

For a good precise assay, we need: (1) sharply defined zone edges so that errors in measurement are minimal, (2) a steep slope of the log-dose line, and (3) a straight log dose-response line (however, this is of lesser importance than the other requirements, as will be explained in Chapter 11).

Attainment of the ideals (1) and (2) must be a compromise; increasing the inoculum will favor sharply defined zone edges but will reduce zone size and steepness of the slope; decreasing the inoculum will increase zone size and steepness of the slope, but if decreased too much will lead to more diffuse zone edges. Prediffusion, at say, 4°C, is comparable to using a lower inoculum level and so increases zone size and slope.

PRINCIPLE OF CALCULATION OF POTENCY ESTIMATE

Although it was shown many years ago that, in general, it is the square of the zone width and not zone diameter unsquared that is directly proportional to logarithm of antibiotic concentration, analysts continue to base their calculations on the latter. No doubt, the reasons are that the procedure became well established before the days of computers. It was seen to be a very satisfactory approximation in most cases and so it was better to avoid the extra steps in computation of the potency estimate. However, it does, on occasions lead to assays being rejected by the statistical tests for curvature. (The importance or unimportance of curvature is discussed in Chapter 11).

The calculation is based on the assumption of two straight parallel lines, one being a plot of responses to standard against the logarithm of dose, the other being a plot of responses to unknown against the logarithm of nominal dose. Doses form a geometrical progression, so that their logarithms form an arithmetical progression. In Europe, the most commonly used assay designs have the same number of dose levels of standard and unknown, i.e., they are *symmetrical* assays.

The calculation may be visualized as in Figure 2.7, which illustrates the meanings of the parameters E, F, b, I, and M by reference to a symmetrical three-dose level assay. Note that this illustration uses logarithms to base 10. (The *European Pharmacopoeia* has, for many years, suggested the use of natural logarithms. Now, however, in the Third Edition, Supplement 2000, it acknowledges that logarithms to base 10 are equally acceptable. [EP 2000, 267]).

The starting points for the calculation are: (1) the logarithm of the standard dose and nominal unknown doses, which are log 1, log 2, and log 4; and (2) S_1, S_2, S_3, U_1, U_2, and U_3, which are the mean responses to low, mid, and high doses of standard and unknown, respectively.

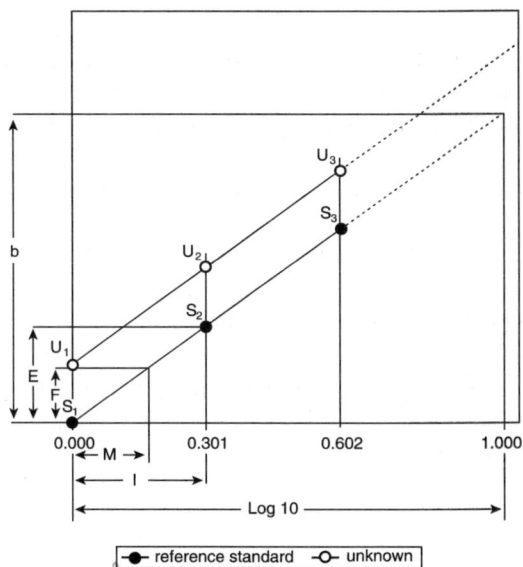

FIGURE 2.7 This diagram uses similar triangles to illustrate the meanings of the parameters *E, F, b, I,* and *M* by reference to a symmetrical three-dose level assay. The horizontal axis denotes dose on a logarithmic scale; the vertical axis indicates the increase in zone size corresponding to an increase in dose.

Thus, on the horizontal scale:
  I is the logarithm of the dose interval (in this case $\log_{10} 2$).
  1.000 is $\log_{10} 10$, which is related to the value of the slope *b*.
  M is the horizontal distance between the two lines and so is the logarithm of the ratio of the potency of the unknown relative to that of the standard.

On the vertical scale:
  $S_1, S_2, S_3, T_1, T_2,$ and T_3 are the mean responses to the three dose levels of standard and unknown, respectively.
  E is the mean difference in response due to the dose interval.
  b (the slope) is the increase in zone size for a tenfold increase in dose.
  F is the mean difference in response due to the difference in potency of the test solutions of the two preparations.

The following are calculated :

  E The difference between mean responses to adjacent dose levels. This is calculated from the differences between mid- and low-dose responses, and between high- and mid-dose responses for both standard and unknown.
  F The difference between the mean response to all unknown doses and the mean response to all standard doses.
  I The logarithm of the ratio between adjacent dose levels.
  b The calculated increase in mean response for a tenfold increase in dose.

M The logarithm of ratio of estimated potency of test solutions for the unknown to that of corresponding standard test solutions.

R The estimated ratio of unknown to standard potency (not shown on the diagram) is obtained as antilogarithm of M.

The calculation of M (in the case of a symmetrical, three-dose level assay) is, thus

$$E = 1/4[(S_3 + U_3) - (S_1 + U_1)] \qquad (2.5)$$

(there are two preparations and two dose intervals);

$$F = 1/3[(U_3 + U_2 + U_1) - (S_3 + S_2 + S_1)] \qquad (2.6)$$

then, considering similar triangles,

$$E/I = b/1.000 \qquad (2.7)$$

$$M/F = I/E = 1/b \qquad (2.8)$$

therefore

$$M = F/b \qquad (2.9)$$

The potency of the unknown test solution relative to that of the reference standard is then obtained as the antilogarithm of M.

For assay designs other than this three-dose level symmetrical design, expressions analogous to Equation (2.5) and Equation (2.6) are given in Appendix 1.

REFERENCES

Cooper, K.E. 1963a. In *Analytical Microbiology*, edited by F.W. Kavanagh. New York: Academic Press, p. 17.

Cooper, K.E. 1963b. In *Analytical Microbiology*, edited by F.W. Kavanagh. New York: Academic Press, p. 51.

Cooper, K.E. 1963 and 1972. In *Analytical Microbiology*, Volume I, Chapter 1, The theory of antibiotic inhibition zones, Volume II, Chapter 2, Diffusion assays edited by F.W. Kavanagh. New York: Academic Press.

Cooper, K.E. and Gillespie, W.A. 1952. The influence of temperature on streptomycin inhibition zones in agar cultures, *J. Gen. Microbiol.*, 7, 1.

Cooper, K.E. and Linton, A.H. 1952. *J. Gen. Microbiol.*, 7, 8.

Cooper, K.E. and Woodman, D. 1946. The diffusion of antiseptics through agar gels, with special reference to the agar cup assay method of estimating the activity of penicillin, *J. Pathol. Bacteriol.*, 58, 75.

EP (*European Pharmacopoeia*). 2000, 267.

Heatley, N.G. 1944. A method for the assay of penicillin, *Biochem. J.*, 38, 61.

Hewitt, W. 1977. In *Microbiological Assay: An Introduction to Quantitative Principles and Evaluation*, New York: Academic Press.

Hewitt, W. 1981. Influence of curvature of response lines in antibiotic agar diffusion assays, *J. Biol. Standardization*, 9, 1.

Humphrey, J.H. and Lightbown, J.W. 1952. A general theory for plate assay of antibiotics with some principal applications, *J. Gen. Microbiol.*, 7, 129.

Lees, K.A. and Tootill, J.P.R. 1955. Microbiological assay on large plates. Part I. General considerations with particular reference to routine assays, *Analyst*, 80, 95.

Mitchison, D.A. and Spicer, C.C. 1949. A method of estimating streptomycin in serum and other body fluids by diffusion through agar enclosed in glass tube, *J. Gen. Microbiol.*, 3, 184.

Vesterdal, J. 1947. *J. Acta Pathol. Microbiol. Scand.*, 24, 273.

3 The Theory and Practice of Tube Assays for Growth-Promoting Substances

INTRODUCTION

When a small inoculum of a microorganism is added to a liquid medium containing an abundance of all the nutrients needed for growth and then that inoculated medium is kept at a suitable temperature, growth proceeds, typically, in accordance with the general pattern illustrated in Figure 3.1. After a short period of slow growth, the lag phase, growth proceeds at a steady rate on a logarithmic scale. This is followed by a period when the population changes only a little due to depletion of the nutrients and possibly also the buildup of waste materials having a bacteriostatic effect. Eventually, the concentration of living cells begins to decline as nutrients become even more depleted and the concentration of bacteriostatic substances becomes greater. These four phases are known respectively as: (a) the lag phase, (b) the log phase, (c) the stationary phase, and (d) the decline phase.

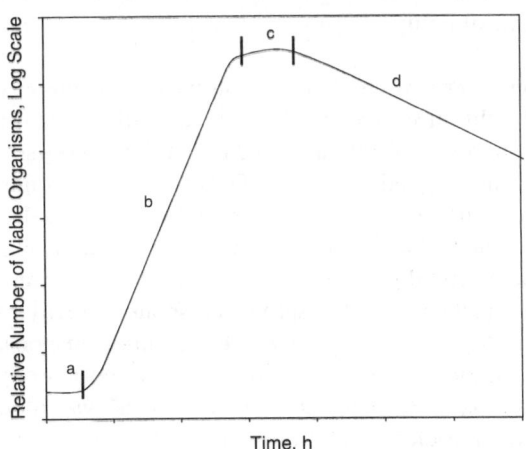

FIGURE 3.1 A graphic representation of the phases of growth and decline of a culture.

In the log phase of growth, the population of living cells doubles in a period of time known as the generation time. The generation time is dependent on the innate characteristics of the organism, the nature of the medium, and the temperature. (As will be seen in Chapter 4, it may also be modified by the presence of growth-inhibiting substances). Generation times may range from about 15 to over 120 minutes for bacteria, from 40 to 180 minutes for yeasts, and from about 90 to 360 minutes for molds. When growth is allowed to proceed as far as the stationary phase, the log phase might, typically, last for about 15 generations. This, naturally, depends on the size of the inoculum. An inoculum of 10,000 viable cells would increase in 15 generations thus:

$$10,000 \times 2^{15} = 3.28 \times 10^8$$

In this chapter, we are concerned especially with the log phase and the way it will be modified in a medium with limited availability of one of the essential nutrients.

The basic concept of tube and flask assays of growth-promoting substances, slope ratio assays, was outlined in Chapter 1. Briefly, the medium contains an abundance of all essential nutrients except one — the substance being assayed (which is absent). Graded additions of the missing essential nutrient lead to gradations in growth on incubation. Now we look at the principles of the assays more closely and see how and why practically obtained results differ from the postulated mathematical basis.

The ideal situation would be a test organism that:

1. Is completely specific in its requirements for a single chemical entity to be assayed
2. Gives a response directly proportional to the dose (or function of dose) of that chemical entity

The actual situation is very different. Many naturally occurring vitamins are families of closely related substances having different activities with respect to the test organism. This is often one of the facts of life that both analysts and recipients of assay results have to understand and accept. Only when assay conditions are properly controlled does the response approach the second of these ideals.

Let us consider now the theoretical aspects of specific response to a single chemical entity, the essential growth-promoting substance. It was stated in Chapter 1 that the ideal dose-response relationship is a response directly proportional to the dose, leading to the slope ratio assay, in which the sample responses plotted against the nominal concentrations of each test solution formed a straight line having a common origin with that of the standard, where response was plotted against actual known concentration of each test solution. This ideal form was shown in Figure 1.2.

Given the maxim that a specific chemical entity enables the growth of cells of lactobacilli, and that those cells produce acid, we have essentially two methods of monitoring growth — either measure the cell growth itself (or some function of cell growth) or measure the acid produced by the cells.

A function of cell growth most frequently employed is turbidity of the resulting cell suspension. The turbidity may be measured by nephelometry, in which light scattered by the suspended cells is measured, or by absorbance of light of defined wavelength. Alternatively, the acid produced in individual flasks may be titrated with alkali.

It has been tacitly assumed that turbidity (however measured) or quantity of acid produced will be directly proportional to concentration of growth-promoting substance. This model (the slope ratio assay) is the basis on which statistical methods of evaluation have been devised.

Methods of statistical analysis have been described by Bliss (1951) and Finney (1978). These methods assess the significance of deviations from the ideal — curvature and failure of regression lines to coincide at zero dose. In fact, assay results frequently display a degree of curvature that varies from slight to severe.

From a consideration of the quantitative effects on growth and acid production when one essential growth factor is present in suboptimal quantities, the possible reasons for curvature of the dose-response line become apparent.

THE MODE OF ACTION OF GROWTH-PROMOTING SUBSTANCES

Two groups of substances required by the lactobacilli for growth are the amino acids and vitamins of the B group. Their roles in promoting growth are quite different. Amino acids are the building blocks for the protein that is a component of the cell. Thus, the amino acids are consumed during cell growth, so that if a particular amino acid were an absolute requirement of the organism, growth would cease when the specific amino acid supply was exhausted.

In contrast, the vitamins of the B group act as coenzymes that are essential for cell metabolism and growth. As catalysts, these substances would play a role in controlling growth rate without being consumed. Based on these precepts, we consider growth rate when limited by suboptimal amounts of growth-promoting substances.

GROWTH LIMITED BY AMINO ACIDS

For convenience, the amino acid that is present in suboptimal amounts will be described as substance A. In an assay in which different tubes contain different quantities of substance A, growth must continue until every molecule of substance A as been consumed. In this theoretical study, assay conditions are so standardized that no factor other than the quantity of substance A in the tube influences growth. It follows that growth will continue for longer in those tubes that contain greater quantities of substance A. The need to ensure that incubation proceeds long enough to permit utilization of all substance A is evident.

During the early and middle part of the incubation period, substance A may be present in sufficient quantity that growth proceeds exponentially or almost so. The rate of growth will diminish toward the end of growth in each individual tube.

The change in rate was described quantitatively by Monod (1949) by the equation:

$$\mu = \frac{\mu_{max}S}{K_s + S} \tag{3.1}$$

where

μ = specific growth rate

μ_{max} = maximum specific growth rate

S = substrate concentration

K_s = a constant that is equal to the substrate concentration when $\mu = 0.5\,\mu_{max}$

Using the Monod equation and making certain realistic assumptions about values of the parameters, theoretical growth curves corresponding to different values of dose (substrate concentration, S) can be drawn. Before looking at such theoretical growth curves, let us see how they are developed.

The values that will be substituted in the Monod equation, although arbitrary, are realistic. Let us suppose:

1. For the growth of one cell of an organism, p molecules of substance A are consumed.
2. All tubes are inoculated with n cells in the logarithmic phase of growth.
3. The generation time when substance A is abundant is G; incubation time will be described in units of G in all cases, whether growth is exponential or not.
4. The quantity of substance A in tubes at dose level 1 is $2^{10} \times np = 1024np$ molecules, which is sufficient to permit an increase of $1024n$ cells/tube; thus, the final cell count in such a tube should be $1025n$.
5. Dose levels 2 and 3 correspond to $2048np$ and $3072np$ molecules/tube, respectively, leading to final cell concentrations of $2049n$ and $3073n$/tube.
6. The constant K_s (which is naturally the same for all dose levels) is 10% of the concentration of substance A at dose level 1 (i.e., it is $102.4np$ molecules/tube).
7. The generation time if A were abundant (i.e., when $\mu = \mu_{max}$) is t minutes.

It follows that at dose level 1, the initial specific growth rate is given by equation (3.1) as

$$\mu = \frac{\mu_{max} \times 1024\,np}{(102.4 + 1024)np} = 0.90909\,\mu_{max}$$

Making the simplifying approximation that the specific growth rate remained constant during the first t minutes of growth, the number of cells would increase from

n to 1.9091n during that period. It follows that the number of molecules of A remaining in the substrate would have dropped from 1024np to $(1,024 - 0.9091)np$ = 1023.1np/tube.

Thus, for the growth period t to 2t, a new value for specific growth rate would be calculated as:

$$\mu = \frac{\mu_{max} \times 1023.1\,np}{(102.4 + 1023.1)\,np} = 0.90902\ \mu_{max}$$

Iterative calculations such as this show that changes in the value of μ are barely perceptible during the first few growth periods and that μ is still greater than 0.9, even after 7t minutes of growth. After this, μ drops at an increasingly rapid rate and it becomes no longer realistic to assume the approximation that it remains constant during t minutes of growth. It is necessary to introduce a modified procedure to recalculate μ after fractional time periods such as 0.1t and even less as the end of growth approaches. With the aid of a computer, it is a simple matter to use the modified procedure for calculations, and so a time period of 0.01t was used to calculate the values in the period when change was greatest.

In addition to dose levels 1, 2, and 3, the calculation was applied to a level of 0.1 representing an adventitious trace of substance A, such as might have been carried over in the inoculum.

The figures calculated for μ, relative cell concentration (RCC), and residual level of critical amino acid in the substrate (RS) are shown in Table 3.1 for dose level 1 only. This demonstrates the dramatic drop in residual level of substance A and in μ as the end of growth approaches. It will be seen that on the basis of the arbitrarily chosen parameters, growth ends after a period of about 12t.

Values for μ and RCC (but not RS) are presented in Table 3.2 for all dose levels. It will be noted that there are large blank areas in this table. This is because these were areas of little change; the omission of these values serves to highlight the areas of greater interest.

From the values for RCC in Table 3.2, the graphs of Figure 3.2 have been derived. In these graphs, the logarithm of RCC is plotted against time. The dotted line represents truly exponential growth, whereas the apparently straight portions of the lines for doses 1, 2, and 3 correspond to nearly exponential growth when the concentration of substance A is barely exerting any growth-limiting effect. In the short, sharply curved parts of the lines, substance A is severely depleted, and in the horizontal parts of the lines, it is virtually absent.

Some practically determined growth curves are shown in Figure 3.3. These are due to Toennies and Shockman (1953). The lines show how growth of *Streptococcus faecalis* is limited by differing levels of L-valine. They differ from the lines of Figure 3.2 in that turbidity continues to increase, although at a reduced rate, well after the nearly exponential phase has ended. Otherwise, they show a remarkable resemblance to the theoretical lines. (It should be noted that in the original publication by Toennies and Shockman, the sharply ascending straight parts of the lines were drawn as being

TABLE 3.1
Theoretical Values for Growth of an Organism in a Medium Having a Limiting Concentration of an Amino Acid

t	μ	RCC	RS
1	0.9091	1.91	1023.1
2	0.9089	3.64	1021.4
3	0.9086	6.96	1018.1
4	0.0981	13.27	1011.9
6	0.9052	48.27	977.4
8	0.8928	174.4	852.9
10	0.8061	606.4	425.7
10.2	0.7782	681.3	359.2
10.4	0.7260	761.9	271.4
10.80	0.5019	925.6	103.2
—	0.5000	—	102.4
10.81	0.3927	929.3	99.4
11.0	0.2830	989.5	40.4
11.2	0.0727	1018.4	8.03
11.4	0.0095	1024.21	0.99
11.5	0.0032	1024.74	0.33
12.0	<0.0001	1025.00	<0.01

Note: μ = specific growth rate; *RCC* = relative cell concentration; *RS* = residual level of the critical amino acid in the substrate; t = time units.

Note: It will be noted that when μ is 0.5, *RS* is 102.4 by definition; no values are given for t and *RCC*. This is because the value of t lies between 10.80 and 10.81 units. The computer program worked in units of $0.1t$.

collinear for doses 2 to 8 µg. They were redrawn as in Figure 3.3 after critical inspection of the raw data with the kind permission of Dr. Shockman).

A series of theoretical dose-response lines based on the same values as used to produce Figure 3.2 but assuming different times for end of incubation is shown in Figure 3.4. These lines demonstrate how curvature at higher dose levels can arise through inadequate incubation. The analogous practical dose-response lines based on the same data as used in Figure 3.3 are shown in Figure 3.5. From these lines, it appears that the response to dose is linear up to 6 µg if incubation is not less than 16 hours.

TABLE 3.2
A Mathematical Illustration of Growth of an Organism in Substrates Having Different Levels of a Limiting Amino Acid

Time Units	Dose Level "Trace"		Dose Level 1		Dose Level 2		Dose Level 3		Exponential Growth
t	μ	RCC	μ	RCC	μ	RCC	μ	RCC	RCC
0.0	0.5000	1.0	0.9091	1.0					
1.0	0.4988	1.5	0.9090	1.9					
2.0	0.4970	2.2	0.9089	3.6					
10.0	0.3589	46.2	0.8098	606.8	0.9274	791.9	0.9567	863.9	1,024
11.0	0.2920	61.3	0.3049	993.2	0.8524	1,503.2	0.9324	1,681.5	2,048
11.2			0.0701	1,020.7	0.7975	1,694.6	0.9202	1,917.1	2,353
11.4			0.0058	1,024.7	0.6686	1,890.2	0.8997	2,181.9	2,702
11.6			0.0004	1,025.0	0.2687	2,035.5	0.8597	2,475.6	3,104
11.8					0.0001	2,049.0	0.7553	2,788.1	3,566
12.0	0.2078	76.7					0.2961	3,045.7	4,096
12.2							0.0002	3,073.0	4,705
14.0	0.0589	97.0							
16.0	0.0094	102.4							
18.0	0.0013	103.3							
20.0	0.0002	103.4							
22.0	0.0001	103.4							

Note: μ = specific growth rate; RCC = relative cell concentration.

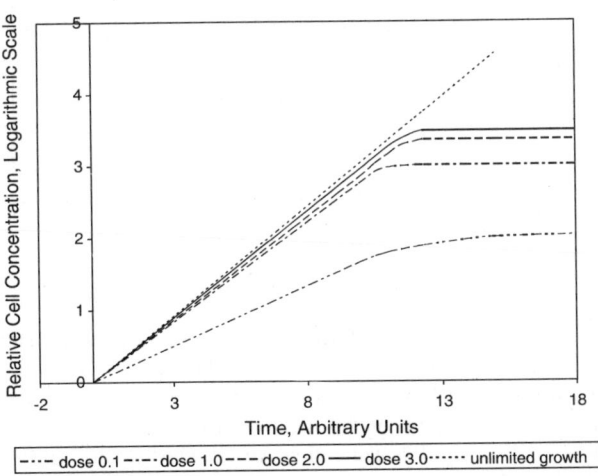

FIGURE 3.2 A representation of cell growth when limited by differing levels of one essential amino acid. These theoretical curves were drawn from the values in Table 3.2 that were calculated using the Monod equation, (3.1).

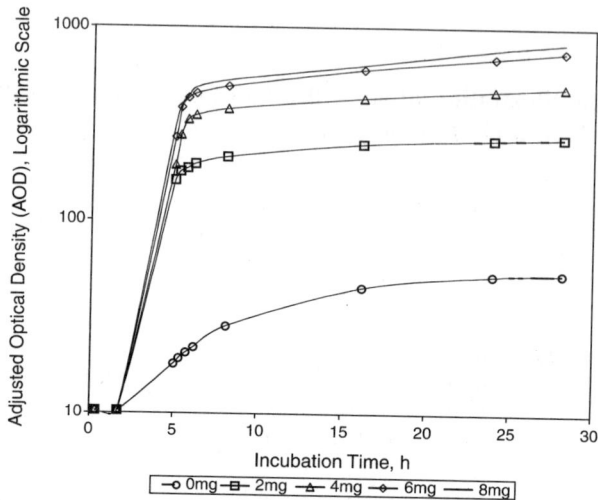

FIGURE 3.3 Practically determined growth curves for *Streptococcus faecalis* (9790) when growth is limited by differing levels of L-valine. Adapted from the work of Toennies and Shockman (1953).

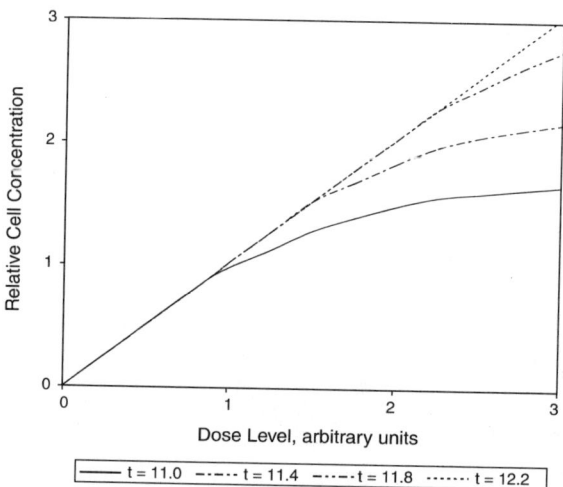

FIGURE 3.4 A representation of the forms of dose-response lines that are predicted at different incubation periods when cell growth is limited by one essential amino acid, *A*. These theoretical curves were drawn from the values in Table 3.2 that were calculated using the Monod equation, (3.1). The length of the incubation period is shown on each line by the number of units of *t* where *t* is the generation period when amino acid *A* is abundant and growth is unlimited.

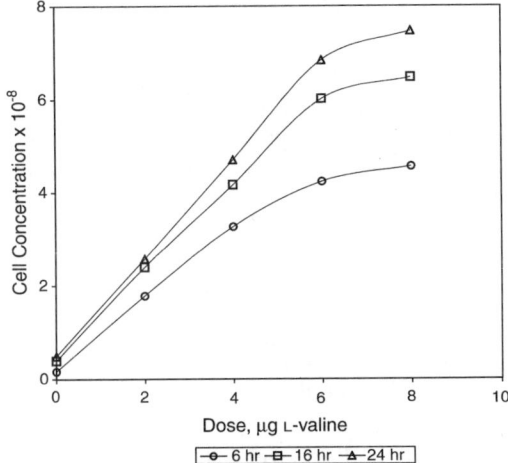

Dose, μg L-valine

FIGURE 3.5 Practically determined dose-response lines after different incubation periods in the assay of L-valine using *Streptococcus faecalis* (9790). These lines are based on the same data of Toennies and Shockman (1953) as were used for Figure 3.3. This demonstrates that for dose levels up to 6 μg an incubation period of 6 hours is inadequate. The dose-response line becomes rectilinear for this dose level at about 16 hours.

GROWTH LIMITED BY VITAMINS

For convenience, the vitamin that is present in suboptimal amount will be described as substance *V*. In the early stages of bacterial growth, even substance *V* will be present in great excess of the requirements of the relatively small numbers of cells, and so it is reasoned that there will be exponential growth not influenced by the varying concentrations of substance *V*. Cell numbers will eventually reach a level at which they are competing for molecules of substance *V* to catalyze their continuing growth. It is at this level that growth will enter an arithmetic phase, in which (assuming no other factors have arisen to exert an influence) growth rate is directly proportional to the concentration of substance *V*.

These precepts enable us to predict the shape of dose-response lines corresponding to different dose levels of substance *V*.

Consider the following model:

1. The inoculum in each tube consists of *n* cells in the logarithmic phase of growth.
2. Substance *V* is present in different tubes at concentrations in the ratio 1:2:3:4 to one another.
3. The concentration of substance *V* at dose level 1 becomes a limiting factor after exactly 12 generations, when the number of cells has become $2^{12} \times n = 4,096n$.

4. The change from logarithmic to arithmetic growth is abrupt (a simplifying assumption).
5. The arithmetic growth rate is taken to be the geometric mean of the rates during the final generation of exponential growth. (Alternatively, the actual rate at the instant that exponential growth ends could be taken. Either assumption leads to the same conclusion, but with slightly different figures.)

It follows from these assumptions that the substance V would become a limiting factor at other dose levels thus:

Dose level 2   13 generations
Dose level 4   14 generations

It can also be calculated that at dose level 3, the rate-limiting factor would become operable at 13.585 generations. The relative cell concentrations thus calculated at different periods of incubation and different levels of V are shown in Table 3.3.

The calculated values from the columns of Table 3.3 are used to plot theoretical growth curves for the organism based on these precepts. These are shown in Figure 3.6, in which the number of organisms is shown on a linear scale, and in Figure 3.7, in which the number is shown on a logarithmic scale. The calculated values from growth periods 20, 30, 40, and 60 are used to plot the dose-response lines that would arise from this growth pattern (Figure 3.8). However, the extent of curvature is

---

TABLE 3.3
Calculated Relative Cell Concentrations after Varying Periods of Incubation When Growth Rate Is Dependent on Four Different Limiting Levels of a Vitamin

	Relative Cell Concentrations			
Growth Period	Dose Level 1	Dose Level 2	Dose Level 3	Dose Level 4
12	**41**	41	41	41
13	61	**82**	82	82
13.585			123	123
14	82	123	148	**164**
20	205	387	517	655
40	614	1188	1746	2294
60	1024	2007	2975	3923

Figures shown in bold indicate the end of exponential growth for the dose level.

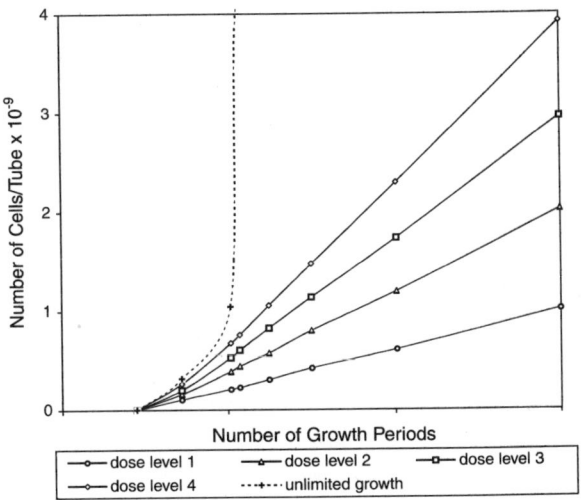

FIGURE 3.6 A series of theoretical curves representing the growth of a test organism according to differing limiting doses of an essential vitamin. These are based on the calculated values of Table 3.3. Growth is shown on an arithmetic scale.

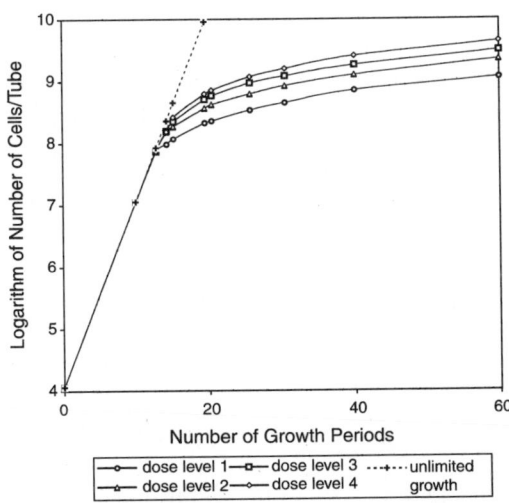

FIGURE 3.7 A series of theoretical curves representing the growth of a test organism according to differing limiting doses of an essential vitamin. These are based on the same values as used for Figure 3.6, but growth is shown on a logarithmic scale.

more apparent from the response ratios of Table 3.4 than from visual inspection of Figure 3.8.

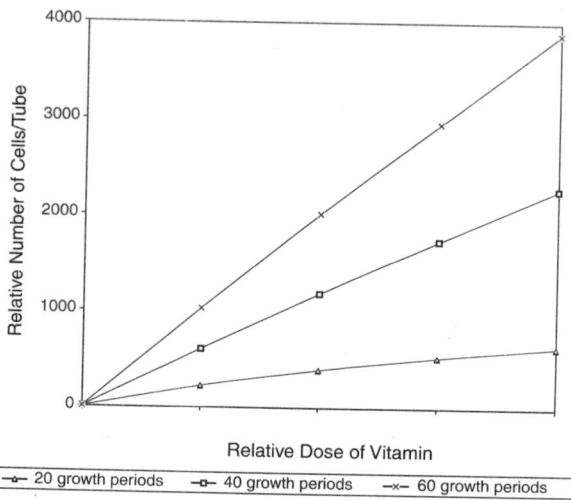

FIGURE 3.8 The form of dose-response lines predicted in a turbidimetric vitamin assay. The lines are based on data such as is presented in Table 3.3. Although it is not very apparent on visual inspection, the line for twenty growth periods is the most strongly curved.

TABLE 3.4
Response Ratios after Varying Periods of Incubation when Growth Rate Is Dependent on Four Different Limiting Levels of a Vitamin

	Response Ratios			
Growth Period	**Dose Level 1**	**Dose Level 2**	**Dose Level 3**	**Dose Level 4**
14	1.00	1.50	1.80	2.00
20	1.00	1.80	2.52	3.20
40	1.00	1.93	2.84	3.74
60	1.00	1.96	2.91	3.84

PRODUCTION OF ACID BY *LACTOBACILLI*

The rate of growth and acid production by lactobacilli was studied by Longsworth and MacInnes (1936). While the purpose of their work was not the development of assay systems, their findings were relevant. The findings of Longsworth and MacInnes were used by Hewitt (1989) to develop theoretical aspects of titrimetric assays. Briefly, it was concluded that the titrimetric measurement system had no advantages over the turbidimetric system so far as linearity was concerned. The titrimetric method appears to be little used nowadays, no doubt because turbidimetric assays are much faster.

It can take up to 3 days for the required concentration of acid to build up. Therefore, titrimetric procedures have been largely replaced by turbidimetric procedures as suitable instrumentation became available. The titrimetric procedure is perhaps mainly of historical interest.

CRITICAL FACTORS IN THE ASSAY OF GROWTH-PROMOTING SUBSTANCES

These are the critical factors and less-critical factors:

Dose levels. These should be high enough to permit adequate growth of the organism so that turbidity can be measured accurately. They should not be so high as to necessitate prolonged incubation to attain linearity of the dose-response line.

Variation in size of inoculum in individual tubes. This is not very serious. Provided that incubation proceeds long enough, small initial differences should not influence final cell concentration.

Incubation temperature. The absolute temperature is not critical, but it should be near the optimum for a good growth rate so that the assay may be concluded in a reasonable time period. Although it is good practice to ensure that temperature is uniform throughout the bath, minor variations in the temperature of individual tubes will not be critical provided that incubation proceeds long enough.

Incubation time. It is vital that this is sufficient to permit a close approach to linearity of the dose-response lines.

Note that this appraisal of what is critical and what is not critical applies to tube assays for growth-promoting substances. Quite different criteria apply to tube assays for growth-inhibiting substances.

OTHER SOURCES OF ERROR

Apart from the general sources of error described in Chapter 6, there are some sources of error peculiar to tube assays for growth-promoting substances:

1. Cleanliness of glassware is of paramount importance. Traces of adsorbed vitamin, antibiotic, or detergent may affect growth in individual tubes, leading to erratic responses.
2. Because of the very low concentration of active substance in test solutions in the case of some assays, adsorption onto glassware during preparation of the test solutions can result in appreciable losses.

REFERENCES

Bliss, C.I. 1951. In *The Vitamins*, Vol II. edited by P. Gyorgy. New York: Academic Press.

Finney, D.J. 1978. *Statistical Method in Biological Assay*, London: Griffin.

Hewitt, W. 1989. In *Theory and Application of Microbiological Assay*, edited by W. Hewitt, and S. Vincent, San Diego: Academic Press, 128–135.

Longsworth, L.G. and MacInnes, D.A. 1936. Bacterial growth at constant pH, Vol. 31, *J. Bacteriol.*, 287.

Monod, J. 1949. The growth of bacterial culture, *Annu. Rev. Microbiol.*, 3, 371–394.

Toennies, G. and Shockman, G.D. 1953. Qualitative Amino Acid Assimilation in Homofermentive Metabolism, *Arch. Biochem. Biophys.*, 45:447.

4 The Theory and Practice of Tube Assays for Growth-Inhibiting Substances

HISTORICAL INTRODUCTION

The turbidimetric method was outlined in Chapter 1, and no further description is needed at this juncture. As with the agar diffusion assay, the need for this method arose in connection with the development of penicillin. Assay methods were required both in the development of production of the antibiotic and for the determination of its concentration in body fluids.

Garrod and Heatley (1944) commented:

> This type of method would seem capable of high precision and might under certain circumstances be the method of choice provided a suitable photometric apparatus were available. Usually a curve relating the light transmitted to the penicillin concentration is constructed from the readings obtained from solutions containing known amounts of penicillin. The concentration of the penicillin in the unknowns is then read from this curve.

Pope (1945) reported that using *Staphylococcus aureus* and a papain digest medium produced a turbidity–penicillin concentration curve that was considerably steeper than with most other media (unpublished data reported in Garrod and Heatley, 1944). The curve, which was sigmoid in shape, had an unusually long and straight central portion, and by extrapolating this back to the axis representing penicillin concentration, a very high degree of accuracy was obtainable.

MEASUREMENT OF RESPONSE

The actual response is the cell mass arising from the growth of the test organism, which varies inversely in accordance with the concentration of the growth-inhibiting substance. Much has been written about the variation in cell size with age of a culture and its effect on optical properties; this is reviewed by Kavanagh (1963). Notwithstanding such variation, it seems reasonable that in a series of tubes having the same initial culture and incubated under identical conditions, the number of cells at the time of termination of growth would be an appropriate response to measure.

In practice, the observed response is normally the turbidity resulting from the cumulative growth of the organism, which may be measured photometrically by light scattered or by light transmitted/absorbed.

Hewitt (1975) compared the responses of three instruments to a series of suspensions of *Staphylococcus aureus*. The organism from an overnight slant was suspended in normal saline, treated with formalin, and diluted to give a suspension having a transmittance of about 50% using a simple absorptiometer at 650 nm (Coleman™ Junior). This suspension (arbitrarily denoted 100) was further diluted to give a series of suspensions of relative concentrations 20, 40, 60, and 80. The absorbancies of these suspensions were measured using two absorptiometers, the Coleman Junior and a Spectronic 20. The results of this work are shown in Figure 4.1, in which the absorbances are plotted against relative cell concentration (RCC). In both cases, the lines are curved due to deviations from the Beer-Lambert law. Deviations from the law would have been less using instruments with monochromators, which give a narrower spread of wavelengths around the peak, that is, a narrower half-band width.

Light scattered by the five suspensions was measured in arbitrary units using a simple instrument, the EEL Nephelometer. Response was directly proportional to RCC. However, the nephelometer had the disadvantage of a slow response; several seconds were needed for the instrument reading to stabilize.

Absorptiometric measurements of cell suspensions and instrumental factors are reviewed by Kavanagh (1963, 1972), who describes modifications to certain

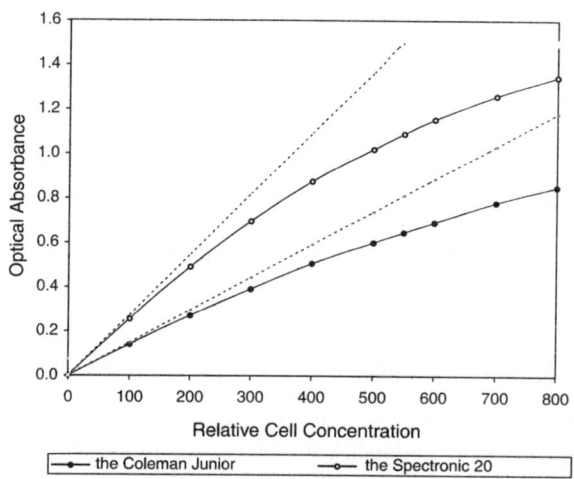

FIGURE 4.1 Calibration of two absorptiometers. The solid lines are the calibration curves relating RCC (*Staphylococcus aureus*) to absorbance at 650 nm. The dashed lines serve to show the extent of deviation from the Beer-Lambert law. For routine transformation of absorbance to RCC it would be convenient to express the relationships in tabular form.

instruments to give a measured response (absorbance) more closely proportional to the true response, which is taken to be cell concentration. This assumption, however, is a deliberate simplification, because cell size may vary according to the age of the culture and the presence of antibiotics, thus affecting optical properties. Calibration curves may be prepared relating cell concentration for any species of organism to absorbance so that assay responses (absorbance) may be converted to RCC.

Kavanagh (1972) stresses the problem of flow birefringence in absorptiometric measurements of suspensions of rod-shaped organisms. Turbulence, after pouring the suspension into the cuvette, causes fluctuation in measurements until the organism has returned to its completely random orientation state. This may take up to 20 sec.

The problem of flow birefringence was overcome in the automated Autoturb® system, which employs a flow cell (A.H. Thomas and Company, No. 9120-NO5) for optical measurements. The orientation of rod-shaped organisms under standard conditions of flow is sufficiently uniform to eliminate fluctuations in measurements.

Hewitt (1977) did not encounter the problem of flow birefringence in nephelometric measurements of suspensions of rod-shaped organisms. This was almost certainly due to the structure of the nephelometer, which collected scattered light over a wide range of angles.

THE FORM OF THE RESPONSE LINE

Consider Figure 1.3 (the postulated form of dose-response line for a growth-inhibiting substance tube assay) showing, for a fixed time of incubation, how total cell concentration varies with concentration of a growth-inhibiting substance. Response (microbial concentration) is plotted against the logarithm of concentration of growth-inhibiting substance. At very low concentrations of growth-inhibiting substance, represented by a to b on the log scale, there is no perceptible inhibition of growth. At concentration e, inhibition is complete and so there is zero growth at concentrations from e to any higher value, f. Intuitively, we would expect these two horizontal parts of the line to be joined by a curved line b to e which would be useful for assay purposes. If the range c to d approximated to a straight line, this would facilitate calculation procedures. This simple logic has indicated in outline the expected shape of the dose-response line. Later in this chapter, a detailed consideration of the kinetics of microbial inhibition will indicate to us more closely the shape that is to be expected. Then we can see how closely the sloping line may approximate to a straight line and how suitable it is for assay purposes.

If the region of the line between concentrations c and d approximates to a straight line, we have the basis of a parallel-line assay. If not, can we use mathematical transformations to convert our observed raw data into a form in which the log dose function-of-response line approaches a straight line in this region?

HISTORICAL DEVELOPMENT OF THE TURBIDIMETRIC METHOD

The development of the turbidimetric assay for growth-inhibiting substances has been empirical. It is interesting to consider the procedures for calculation of potency estimates from observed data that have been reported or suggested both officially and unofficially over the years. Most of these suggestions have emanated from the United States, where there has been more interest in turbidimetric assays than there has been in Europe:

1. Plot *transmittance* against *dose*. This was the procedure used in the early work reported by Garrod and Heatley (1944).
2. Average the *transmittances* for each treatment, and plot their means on semilog paper with *doses* on the log scale and *transmittances* (means) on the arithmetic scale. Draw a smooth curve through the points. If the points appear to represent a straight line, calculate the ideal high and low responses. This procedure appeared in the *United States Pharmacopeia (USP) 16* (1960) but not in later editions.
3. The *USP 17* (1965) suggested plotting (100% transmittance) against *log dose*. *Transmittance* could be changed to *absorbance* and, if a linear response is not obtained when expressed in this way, the response could be linearized by expressing absorbance in terms of percent reduction in growth, and thence converting to *probits*. (Probits will be considered in the "Linearization of Sigmoid Curves" section of this chapter.)
4. Plot *absorbance* against *log dose*. This method has been recommended for most assays in the United States Code of Federal Regulations (CFR).
5. Kavanagh (1963) suggested the use of *relative cell concentration (RCC)* rather than a direct measurement of optical properties. *RCC* was based on a practically determined relationship between actual *RCCs* of a defined microorganism and observed optical absorbance. It was implicit that *RCC* should be plotted against *logarithm of dose*.
6. The *USP 18* (1970) suggested either three dose levels each of standard and sample or a five-dose level standard curve with one dose level of sample. Although the pharmacopoeia did not specify what function of response was to be measured, it was implicit that optical *absorbance* should be used and plotted against logarithm of dose.
7. The second edition of the *European Pharmacopoeia (EP)* (1993) required dose-response lines for standard and sample to be both linear and parallel within a defined probability level. It went on to say that the method of calculation would depend on the optical apparatus used. A suitable transformation of response should be found so as to ensure linearity of the lines. Although the pharmacopoeia did not suggest any specific transformation, it was implicit that this could be the probit or other transformation that is described in the "Linearization of Sigmoid Curves" section of this chapter.

The use of light transmittance directly has no logical basis and generally leads to strongly curved lines, although it is noted (as mentioned earlier) that Pope claimed for specific conditions, "an unusually long and straight central portion" when plotting transmittance against dose (as reported in Garrod and Heatley, 1944).

Optical absorbance may be approximately proportional to cell concentration and thus display the sigmoid shape that is expected. However, assay conditions are normally such that the entire curve is not displayed. This is entirely logical because (referring to Figure 1.3) the regions corresponding to concentrations *a* to *b* and *e* to *f* would serve no purpose whatsoever, and the ranges *b* to *c* and *d* to *e* would be of lesser value than the range *c* to *d*.

The results of applying some of these procedures may be seen in the examples that follow.

Example 4.1

In a tube assay of streptomycin using *Klebsiella pneumoniae*, a plot of percent transmission of light at 530 nm vs. dose is shown in Figure 4.2.

The eight-point standard curve is distinctly sigmoid; the difficulty of fitting a smooth curve to the points is clear. The three dose levels of unknown have not been well chosen because, on visual inspection of Figure 4.2, they appear to represent a sample of half

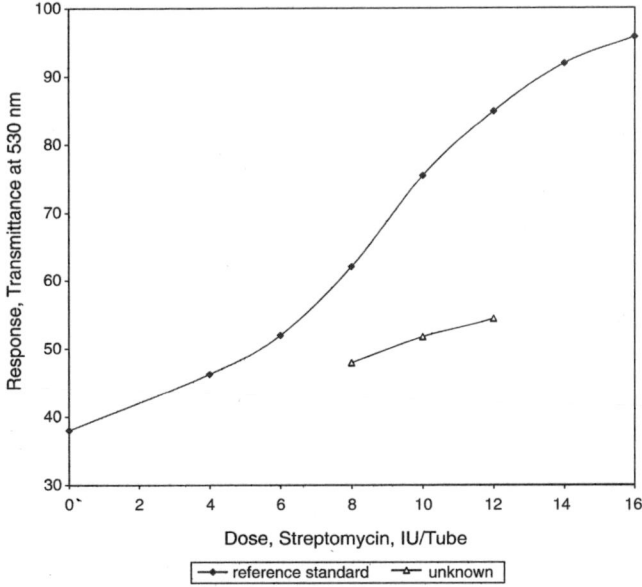

FIGURE 4.2 Dose-response lines representing a turbidimetric assay of streptomycin using *Klebsiella pneumoniae*. Transmittance at 530 nm is plotted against dose. The unknown is about half the potency of the standard. If the line had been plotted at nominal potencies of 4, 5, and 6 IU/tube instead of 8, 10, and 12 IU/tube, there would have been close coincidence with the line for standard.

the potency of the standard. If the unknown responses were replotted at nominal potencies of 4, 5, and 6 instead of 8, 10, and 12, the line would coincide closely with the standard curve. This suggests a valid assay, although coincidence is in the lower, strongly curved portion of the line. Potency could be estimated by interpolation from the standard curve at three points and by taking the mean.

The raw data have been transformed and plotted as absorbance vs. log dose in Figure 4.3. This has yielded a line that is still sigmoid but less strongly curved. Again, if the sample points were replotted as log of nominal potencies 4, 5, and 6, the line would more closely coincide with that of the standard, and although it would not be in the central part of the line, which has the best slope, it would probably be close enough to straight to warrant use of a parallel-line calculation based on the three closest points of the standard curve.

Example 4.2

This is an example of a tube assay for tetracycline using *Staphylococcus aureus*. The six dose levels of both standard and sample form a geometrical progression in the ratio 3:2. Potencies of sample test solutions are reasonably close to those of corresponding standards. The plot of absorbance at 530 nm vs. dose on a log scale, Figure 4.4, shows two somewhat sigmoid curves. Inspection of the lines suggests that there is no doubt that potency could be estimated using only responses to the two middle dose levels (0.18 and 0.27 IU/tube) in a 2 + 2 calculation. It also seems possible that responses to the four middle dose levels (0.12 to 0.405 IU/tube) could be used in a 4 + 4 calculation. However, such procedures would be wasteful in effort, utilizing only 33% and 67% of the observed data, respectively.

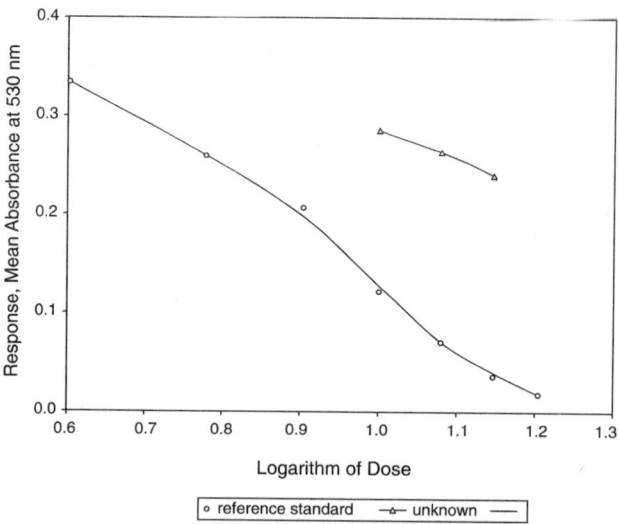

FIGURE 4.3 Dose response lines representing a turbidimetric assay of streptomycin using *Klebsiella pneumoniae*. This is from the same raw data as Figure 4.2, but absorbance is plotted against logarithm of dose. Curvature is somewhat less marked than in Figure 4.2.

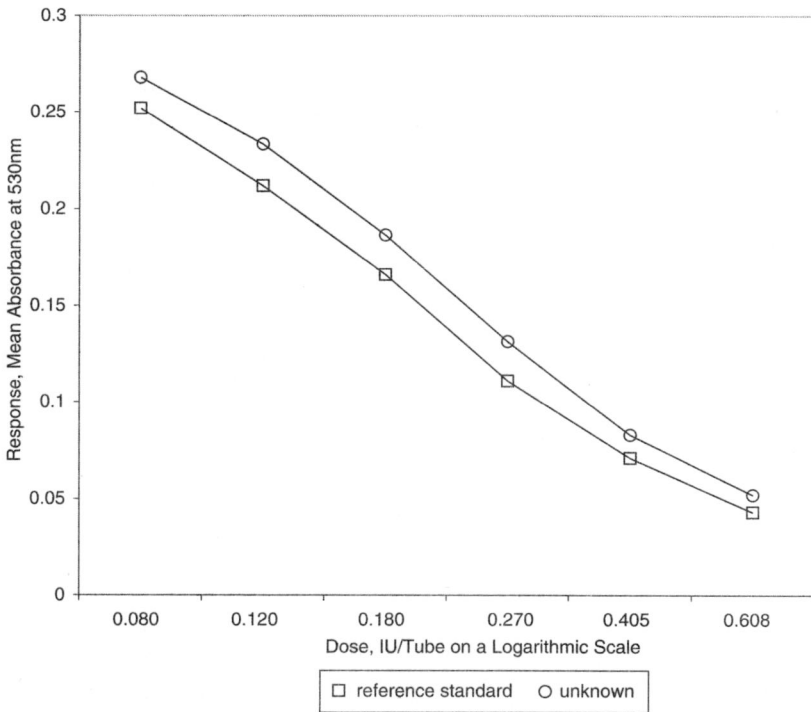

FIGURE 4.4 Dose-response lines representing a turbidimetric assay of tetracycline using *Staphylococcus aureus*. This is a symmetrical six-dose level assay with a 3:2 dose ratio between adjacent doses. Absorbance is plotted against logarithm of dose.

LINEARIZATION OF SIGMOID CURVES

The need to transform sigmoid response lines into a more convenient form for arithmetical manipulation arose in connection with *macro*biological assays. Procedures for linearization were proposed independently by Hemmingsen (1933) and Gaddum (1933). They used the normal distribution curve in its cumulative form to manipulate quantal responses into a more manageable form. The two forms of the normal distribution curve are shown in Figures 4.5a and b. A quantal response is an all-or-none effect, such as the death of an animal in a group. For example, in assaying insulin by the mouse method, suppose the mice were divided into four groups of 50 mice. (This is probably a larger number than would be used in practice. For the purpose of this illustration, it is the smallest number of which both 16 and 50% are whole numbers.) Each mouse in a group would receive the same dose of a test solution. The four test solutions would be low-dose standard, high-dose standard, low-dose unknown, and high-dose unknown. A positive response is either convulsion or death of a mouse. Possible positive responses in a group range from zero to 50. If there were 25 positive responses in one group (50%), that would correspond to zero deviation from the mean of all possible responses. This is shown graphically in Figure 4.5b, where the value 50% for cumulative frequency corresponds to zero deviation from the mean on the

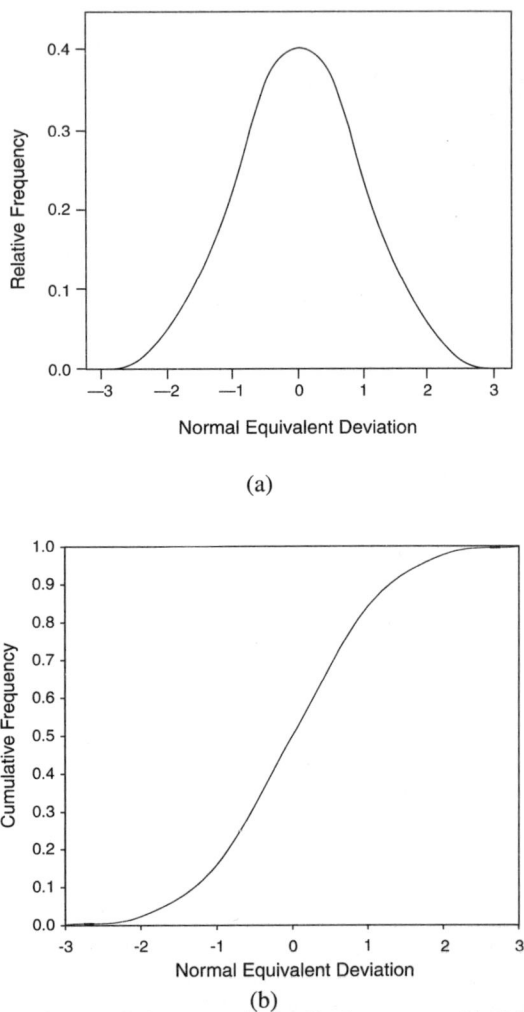

FIGURE 4.5 The two forms of the normal distribution curve: (a) the normal frequency distribution curve and (b) the cumulative normal frequency distribution curve.

horizontal axis. Similarly, a group in which there were eight (16%) positive responses corresponds to one (negative) normal equivalent deviation from the mean. Plotting the normal equivalent deviation corresponding to percent positive responses in a group against the logarithm of the dose received by that group should yield a straight line.

To avoid the use of negative numbers, Bliss (1934b) added five to all values of the normal equivalent deviation and gave the new values the name *probability unit* which was shortened to *probit*. Rather than read from a graph, it is more convenient to refer to tables of probits such as those in *Statistical Tables* by Fisher and Yates (1963).

Although in microbiological assays we are not dealing with quantal responses, it is quite valid to use the probit transformation if it has the desired effect.

An alternative procedure described for processing quantal data is the angular transformation, which is defined as

$$p = \sin^2\phi \qquad (4.1)$$

where

p = the proportionate response (percent of subjects reacting)
ϕ = is an angle between $0°$ and $90°$

Berkson (1944) described the "logit transformation," which was intended for use with quantitative as distinct from quantal responses. Berkson defined the logit y as

$$y = \ln(p/q) \qquad (4.2)$$

where

p = the proportionate response
$q = 1 - p$

However, Fisher and Yates (1963) in effect used the same transformation with the same name but differing by a factor of 2. They defined their logit z as

$$z = 0.5 \ln(p/q) \qquad (4.3)$$

The three transformations have an almost identical effect at proportionate responses in the range 20 to 80%, as may be seen from Figure 4.6.

When such transformations are used, there is a complication in that observations at the extremes of proportionate response are of lesser reliability than those in the central region. To make allowance for this, weighting factors are introduced in the cases of the probit and logit transformation. These weighting factors are obtained from statistical tables such as those of Fisher and Yates (1963).

Such transformations appear to be little used in *micro*biological assay; we have not found them very successful, and so they are not discussed further here.

THE QUANTITATIVE THEORY OF MICROBIAL GROWTH AND INHIBITION

General experience of turbidimetric assays shows that when some measure of turbidity — absorptiometric or nephelometric — is plotted against logarithm of a dose of growth-inhibiting substance over a sufficiently great range of concentrations, the resulting line approximates to part of a sigmoid curve, as is suggested by Figure 1.3.

Various mechanisms could be postulated to explain this sort of curve:

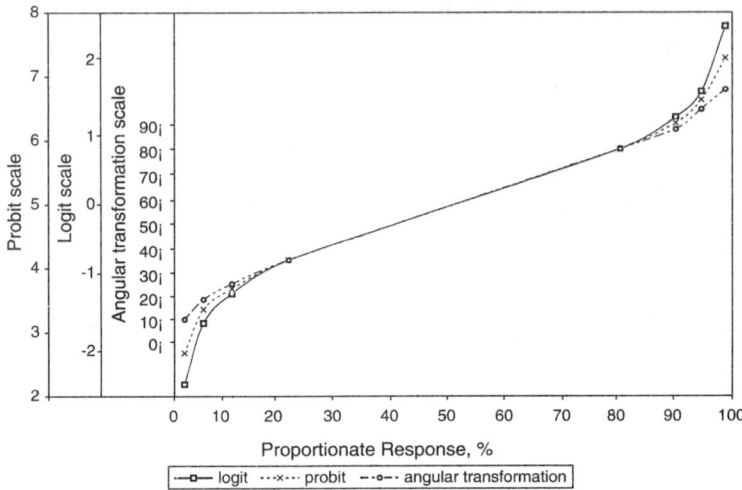

FIGURE 4.6 A comparison of the probit, logit, and angular transformation of response. The three scales of the y axis have been selected to demonstrate the almost indistiguishable effect of these transformations at values of x between 20 and 80%.

1. The duration of the lag phase is increased with increasing concentration of growth-inhibiting substance.
2. The generation time is increased with increasing concentration of growth-inhibiting substance.
3. The growth-inhibiting substance kills a proportion of the organisms, which is related to its concentration, and permits the remainder to grow just as if the growth-inhibiting substance were not present.

Edward Garrett and his team at the University of Florida published a series of papers on the kinetics and mechanisms of actions of antibiotics on microorganisms (Garrett and Miller, 1965). In the cases they studied, it was shown that inhibition was by mechanism 2.

Kavanagh (1968) has assessed that in the majority of cases of growth inhibition that have been studied, mechanism 2 is the main effect. He suggests that this is an appropriate representation of the responses to tylosin, penicillin, and erythromycin, as well as tetracycline and chloramphenicol (which were featured in the work of the Garrett team).

Garrett and Miller (1965) studied the effect of varying concentrations of both tetracycline and chloramphenicol on the growth of *Escherichia coli* in a liquid medium. The purpose of this work was not to provide a theoretical basis for turbidimetric assays; it was part of a broader study of structure/activity relationships. However, their findings are relevant to our subject, and so they are illustrated here by the case of tetracycline.

An *Escherichia coli* culture in the logarithmic phase of growth was used to inoculate a series of loosely capped 125-ml conical flasks, each containing 20 ml of broth. The quantity of inoculum was such as to give an initial cell concentration

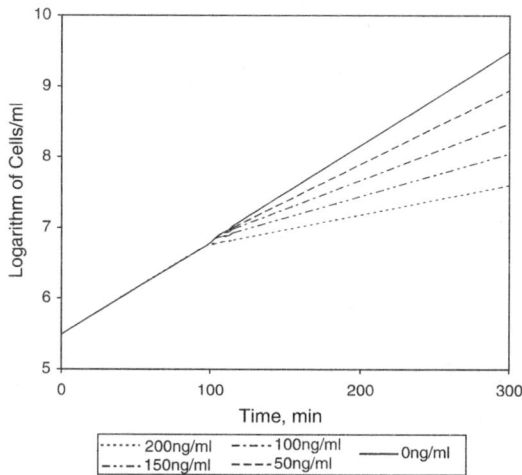

FIGURE 4.7 The influence of varying concentrations of tetracycline on the growth of *Escherichia coli*. At tetracycline hydrochloride concentrations of 50 to 200 ng/ml (molarity about 1 to 4 × 10⁻⁷), total and viable counts were almost identical and are represented by a single line for each concentration.

of 10⁶/ml. The flasks were then incubated for 90 minutes at 37.5°C in a large water bath equipped with a shaker. At the end of the 90-minute period, tetracycline was added to flasks to give concentrations of 0.0, 1.04, 2.08, 3.12, and 4.16 × 10⁻⁷ *M*. These correspond to 0, 50, 100, 150, and 200 ng/ml of the hydrochloride.

Incubation was continued for 6 hours. During that period, total cell counts (using a Coulter counter) and viable cell counts (plate count) were made. The results of this work are expressed graphically in Figure 4.7. It was shown that in this range (50 to 200 ng/ml), there was close coincidence between total and viable counts. The graph in Garrett and Miller's original publication showed the observed points; it is a very complicated picture with 10 different symbols corresponding to total and viable counts for zero-dose and four dose-levels. For this reason, the observed points are not shown in Figure 4.7. Suffice it to say that only moderate scatter of observations about the lines was observed. As seen from Figure 4.7, growth was exponential, but at reducing rates with increasing concentration of tetracycline.

Generation-rate constants, k (where k is greater than zero), were calculated from

$$N = N_0 e^{kt} \tag{4.4}$$

where

> t = the duration of incubation (seconds)
> k = the generation-rate constant
> N_0 = the initial cell concentration
> N = the cell concentration at incubation time t
> e = the base of natural logarithms, 2.71828

The calculation is illustrated by reference to the 100 ng/ml line.

From the graph, the increase in log cell count during 200 minutes is 1.536. This corresponds to a 34.356-fold increase. Substituting in Equation (4.1)

$$34.356 \, N_0 = N_0 e^{kt}$$

$$\ln 34.356 = k \times 200 \times 60$$

$$3.5368 = k \times 12,000$$

$$k = 2.95 \times 10^{-4}$$

The generation-rate constant k (based on time in seconds) is related to generation time G (minutes) by the equations:

$$G = \ln 2/60k \qquad\qquad (4.5a)$$

or

$$k = \ln 2/60G \qquad\qquad (4.5b)$$

Figure 4.8 shows the linear relationship between generation-rate constant and concentration of tetracycline.

The general equation for the line is

$$k = k_0 - k_A A \qquad\qquad (4.6)$$

where

$\quad$ k = the apparent generation-rate constant in the presence of the antibiotic
$\quad$ k_0 = the generation-rate constant in the absence of the antibiotic
$\quad$ k_A = the inhibitory-rate constant for the antibiotic
$\quad$ A = the molarity of the antibiotic

The specific equation representing the graph, Figure 4.8, is

$$k = 4.6 \times 10^{-4} - (7.93 \times 10^2)A$$

To illustrate this, the value of k is calculated for tetracycline molarity 4.12×10^{-7} (200 ng/ml).

$$k = \left(4.6 \times 10^{-4}\right) - \left(793 \times 4.12 \times 10^{-7}\right)$$

$$= \left(4.6 \times 10^{-4}\right) - \left(3.267 \times 10^{-4}\right)$$

$$= 1.333 \times 10^{-4}$$

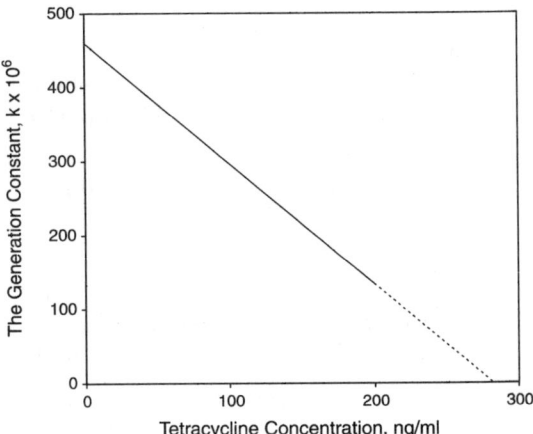

FIGURE 4.8 Illustration of the rectilinear relationship between generation rate constant for *Escherichia coli*, *k*, and concentration of tetracycline in the range 0 to 200 ng/ml. Values of *k* were calculated from the equation

$$k = 4.6 \times 10^{-4} - (7.93 \times 10^{2})\, A$$

where *A* is the molarity of the antibiotic. A molarity of 2.06×10^{-7} corresponds to a tetracycline hydrochloride concentration of 100 ng/ml. Values of *k* calculated for tetracycline concentrations of greater than 200 ng/ml are based on extrapolation. The relationship breaks down as concentration of tetracycline approaches 278 ng/ml and *k* approaches zero.

The generation times and generation-rate constants obtained from these data are shown in Table 4.1. The values for concentrations of tetracycline hydrochloride greater than 200 ng/ml are, of course, based on extrapolation, which may not be valid. The relationship breaks down at concentrations greater than 278 ng/ml as *k* approaches zero.

Using the generation-time constants of Table 4.1, a series of values of final cell counts was calculated for incubation times of 200 and 240 minutes. The calculation is illustrated for a log-dose value of 1.4 and an incubation period of 240 minutes; log-dose 1.4 corresponds to 25.12 ng/ml and a molarity of 0.522×10^{-7}

then

$$k = 4.6 \times 10^{-4} - (793 \times 0.522 \times 10^{-7}) = 4.186 \times 10^{-4}$$

The number of cells arising from one cell after 240 minutes (14,400 sec) incubation is

$$[1 + (4.186 \times 10^{-4})]^{14,400} = [1.0004186]^{14,400} = 414$$

TABLE 4.1
The Generation Times and Generation Rate Constants for *Escherichia coli* in a Liquid Medium at 37.5°C in the Presence of a Range of Concentrations of Tetracycline Hydrochloride

Tetracycline Hydrochloride Concentration ng/ml	Generation Rate Constant k, sec^{-1}	Generation Time, G, min
0	460×10^{-6}	25.1
50	378×10^{-6}	30.6
100	295×10^{-6}	39.2
150	213×10^{-6}	54.2
200	130×10^{-6}	88.8
250	48×10^{-6}	240.7
270	15×10^{-6}	770.2
278	1.6×10^{-6}	7220.0

This and similarly obtained values for *RCC* are plotted against logarithm of tetracycline hydrochloride concentration for both 200 and 240 minutes in Figure 4.9.

It will be seen that the curves of Figure 4.9 do approximate to the form postulated in Figure 1.3. However, they do not have the symmetry of the cumulative normal distribution curve (Figure 4.5b). This absence of symmetry is, no doubt, the reason that probit, logit, and angular transformations have not proven very useful.

Figure 4.10 again shows the curve for 240 minutes incubation (now regarded as the standard), together with a 240-minute curve representing an unknown of higher actual concentration of antibiotic, but which has been plotted at the same nominal concentrations. It is seen that the almost-straight central portions of the curves are parallel; this suggests a basis for a parallel-line assay.

A PRACTICALLY DETERMINED LOG DOSE-RESPONSE CURVE

Preparatory work for the development of a turbidimetric assay for gramicidin was carried out in a U.K. pharmaceutical company in 1994. The test organism was *Enterococcus durans*. The log dose-response line was investigated over a range of doses from 0.0001 to 0.1 µg/tube, representing an overall ratio of 1000:1.

A graph representing this relationship is shown in Figure 4.11, where each point is the mean of four observed responses. This clearly corresponds to a large portion of the overall log dose-response line as predicted in Figure 1.3 and Figure 4.9.

The graph was used to select suitable dose levels for a routine assay. The dose range 0.0015 to 0.0060 µg/tube corresponded to the central, almost-straight part of the line.

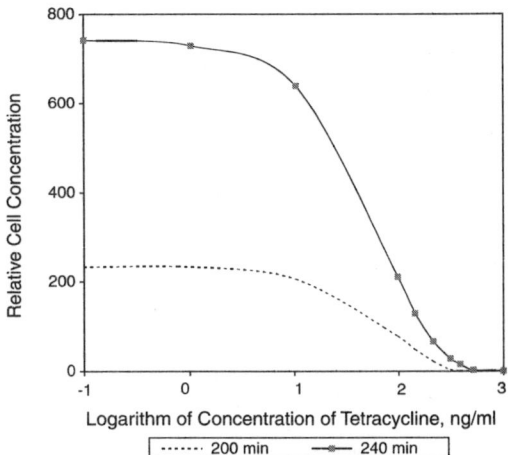

FIGURE 4.9 The expected (calculated) form of the log dose-response curves for a turbidimetric assay using *Escherichia coli* and tetracycline. Plots of *RCC* against logarithm of antibiotic concentration are shown for growth periods of 200 and 240 minutes. Values for the *RCC* were calculated from the growth rate constants, k, as presented in Table 4.1.

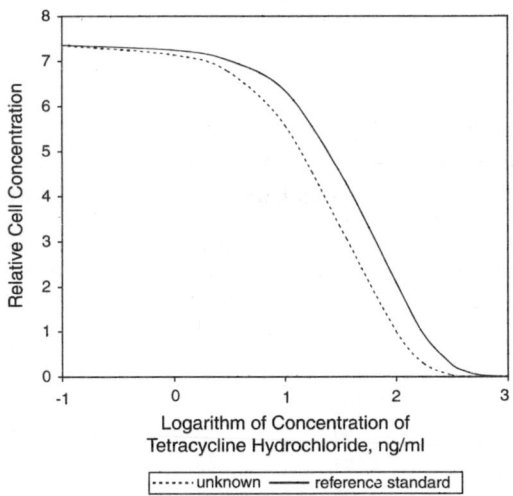

FIGURE 4.10 The same data as in the 240-minute curve from Figure 4.9 representing the reference standard, but with the addition of a curve for an unknown having a potency 58% greater than that of the reference standard.

FACTORS AFFECTING FINAL CELL COUNT

Changes in the generation period and lag time with temperature were studied by Cooper (1963) in the case of *Staphylococcus aureus* grown in broth. His findings, together with the calculated generation-rate constants, are shown in Table 4.2.

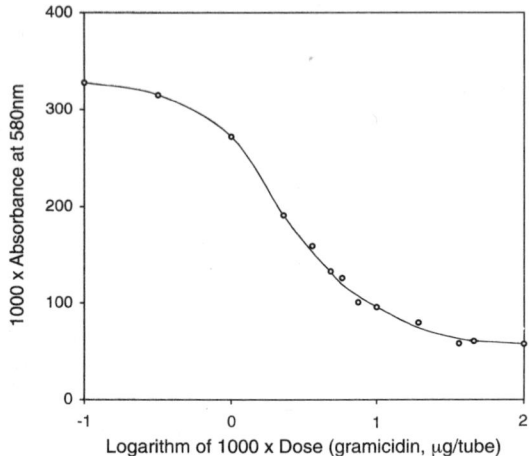

FIGURE 4.11 A practically determined log dose-response curve: gramicidin with test organism *Enterococcus durans*.

TABLE 4.2
Changes in the Generation Period, Generation Constant, and Lag Time with Temperature of a *Staphylococcus aureus* Culture Grown in Broth

Temperature, °C	Lag Time L, min	Generation Period G, min	Generation-Rate Constant k, sec × 10^{-4}
22	180	89	1.298
25	120	64	1.805
26	102	52	2.222
27	90	50	2.310
27.5	84	48	2.407
30	66	38	3.040
32.5	57	33	3.501
35	50	29	3.984
35.5	48	29	3.984
37	45	28	4.126
38.5	42	27	4.279
40	39	27	4.279
41.5	37	26	4.443
42	36	25	4.621

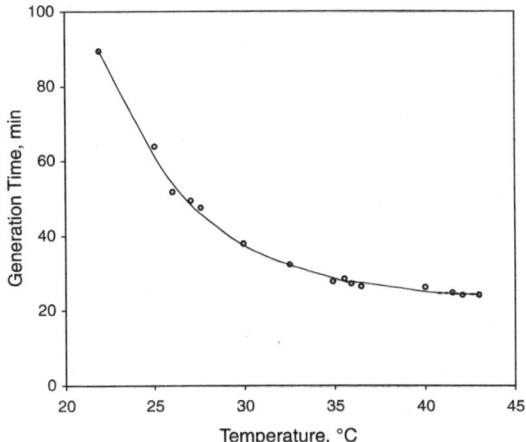

FIGURE 4.12 A plot of generation time G vs. temperature. The values plotted were obtained from Table 4.2.

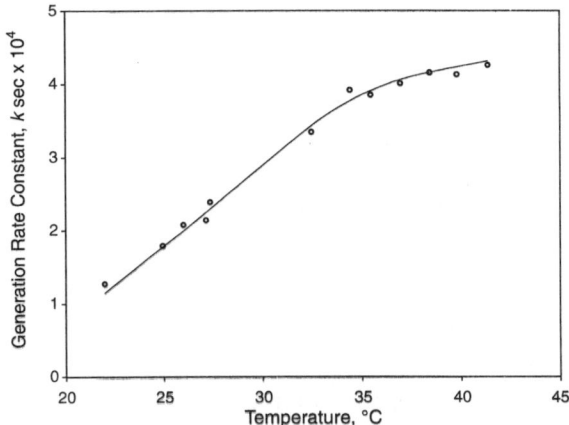

FIGURE 4.13 A plot of generation rate constant k vs. temperature. The values plotted were obtained from Table 4.2.

Using the values from Table 4.2, graphs were plotted of generation period G vs. temperature in Figure 4.12 and generation-rate constant k vs. temperature in Figure 4.13. Although the former is more illustrative of the effect of temperature, the latter is more useful because it shows that k increases linearly up to a temperature of about 33°C. Extrapolating to lower temperatures, at 17.3°C, $k = 0$. The straight-line part of the graph is represented by the equation

$$k = 0.00002338 \, (T - 17.3)$$

where T is temperature, °C.

This equation was used to calculate the best estimates of k for any temperature in the linear range. It was thus possible to calculate the effect of incubation temperature on final cell count in a tube.

It is obvious that both time and temperature must influence final cell concentration. It is important to know the magnitude of these influences.

THE INFLUENCE OF TEMPERATURE

Using the best estimates of k for various temperatures, the increase in the value of N/N_0, the ratio of final to initial cell concentration, could be calculated for various incubation periods by

$$N/N_0 = (1 + k)^{60t} \tag{4.7}$$

where t is the period of incubation (logarithmic phase) in minutes.

Incubation at 30°C for 180 minutes gave

$$k = 0.00002338 (30.0 - 17.3) = 0.000296926$$

and

$$N/N_0 = (1.000296926)^{60 \times 180} = 24.69$$

Similarly, incubation at 30.1°C for 180 minutes gave

$$k = 0.00002338 (30.1 - 17.3) = 0.000299264$$

and

$$N/N_0 = (1.000299264)^{60 \times 180} = 25.32$$

Thus the increase in cell concentration for an increase of 0.1°C was given by

$$[(25.32 - 24.69)/24.69] \times 100 = 2.6\%$$

Similar calculations for an increase of 0.2°C for both 180 and 240 minutes incubation in the log phase gave the values assembled in Table 4.3. The percentage increase in final cell concentration arising from a 0.1°C increment in temperature is the same, whether comparing 30.0 and 30.1°C or 35.0 and 35.1°C.

THE INFLUENCE OF TIME

The effect of substantial differences in incubation time may be seen by comparing the values of N/N_0 at the same incubation temperature for 180 and 240 minutes in Table 4.3. A difference of one hour has caused a threefold increase in cell concentration in this rather slow-growing culture (for which $G = 37.5$ minutes at 30°C).

TABLE 4.3
Variations in Final Cell Count Depending on Incubation Time and Temperature in the Absence of a Growth-Inhibiting Substance

Temperature °C	N/N_0 after 180 min Incubation	Percent Increase	N/N_0 after 240 min Incubation	Percent Increase
30.0	24.7		71.9	
30.1	25.3	2.6	74.4	2.6
30.2	26.0	5.2	76.9	5.2
35.0	87.2		386.8	
35.1	89.5	2.6	400.0	3.4
35.2	91.7	5.2	413.7	7.0

Note: The percentage increase in final cell concentration arising from a 0.1°C increment in temperature is the same whether comparing 30.0 and 30.1°C or 35.0 and 35.1°C.

A simple way of looking at the potential for increase is to consider a culture having a typical generation time of 20 minutes. In one hour, the cell count would increase eightfold. It is quite clear why as little as 30 minutes can make all the difference between optical properties of tubes being readable or not readable.

There is a potential for bias in estimated potency if different tubes in an assay are not incubated for exactly the same period. Suppose that in a set of tubes that comprise an assay, all tubes are of identical shape and thickness of glass. They are all placed in the incubation bath at the same instant; the bath is well stirred so that all tubes reach incubation temperature at the same rate. Incubation time differs, however, as the organism is not killed at the same instant in all tubes.

Consider then how the final cell concentration would differ in tubes having the same concentration of growth-inhibiting substance but incubation periods differing by one minute.

Suppose that for a particular growth-inhibiting substance concentration, the following parameters apply:

$$\text{growth rate constant } k = 0.000770163$$

$$\text{equivalent to generation time } G = 15.0 \text{ minutes}$$

further suppose that

$$\text{the intended incubation period in logarithmic phase} = 180 \text{ minutes}$$

but due to delay in ending incubation,

the period for one tube = 181 minutes

The calculations using Equation (4.7) are

$$N/N_0 = (1.000770163)^{60 \times 180} = 4080.07$$

$$N/N_0 = (1.000770163)^{60 \times 181} = 4277.59$$

An increase of $[(4277.59/4080.07) - 1] \times 100 = 4.84\%$

However, at longer generation times, the effect is less dramatic, as illustrated in Table 4.4, which is based on the growth-rate coefficients derived from the data of Garrett and Miller in Table 4.1 for tetracycline concentrations of 50, 100, and 200 ng/ml.

SUMMARY AND CONCLUSIONS

The problems of nonlinearity of responses in growth-inhibiting substance turbidimetric assays, as well as means of alleviating them, have been seen. When the

TABLE 4.4
Variations in Final Cell Count Depending on Incubation Time and Temperature in the Presence of Varying Concentrations of a Growth-Inhibiting Substance

Tetracycline ng/ml	Growth Rate Constant k	Generation Period, min	Growth Period, min	N/N_0	Percent Increase in N/N_0 in 1 min
0	0.000479	24.1	180	176.3	
			181	181.4	2.9
50	0.000378	30.6	180	59.2	
			181	60.6	2.3
100	0.000295	39.2	180	24.18	
			181	24.61	1.8
200	0.000130	88.9	180	4.04	
			181	4.10	0.8

Note: These figures are based on the growth-rate coefficients derived from the data of Garrett and Miller in Table 4.1 for tetracycline hydrochloride concentrations of 50, 100, and 200 ng/ml. It is seen that at longer generation periods the effect of change in incubation time is less dramatic than was seen in Table 4.3.

responses or modified responses are such that standard and sample response lines are parallel, then a symmetrical assay design is desirable.

Critical factors in determining the final cell concentration in an individual tube may be summarized as:

Initial cell concentration, because on theoretical grounds the final cell concentration should be directly proportional to the initial concentration
Temperature of incubation
Duration of incubation

The influence of temperature and duration of incubation have been demonstrated using only one microorganism, *Staphylococcus aureus*. It is suggested that the conclusions reached in that case are indicative of the extent of such variations in time and temperature in the case of other microorganisms.

REFERENCES

Berkson, J. 1944. *J. Amer. Statist. Assoc.*, 39, 357.

Bliss, C.I. 1934a. *Science*, 79, 38.

Bliss, C.I. 1934b. *Science*, 79, 409.

Code of Federal Regulations.

Cooper, K.E. 1963. In *Analytical Microbiology*, Vol. 1, edited by F.W. Kavanagh, New York: Academic Press.

European Pharmacopoeia. 1993. Vol. II.

Fisher, R.A. and Yates, F. 1963. *Statistical Tables for Use in Biological, Agricultural and Medical Research*, 6th ed., London: Longman.

Gaddum, J.H. 1933. Medical Research Council. Special Report Series No. 183.

Garrett, E.R. and Miller, G.H. 1965. Kinetic and mechanisms of action of antibiotics on microorganisms III. *J. Pharm. Sci.*, 54, 427.

Garrod, L.P. and Heatley, N.G. 1944. Bacteriological methods in connection with penicillin treatment, *Br. J. Surg.* XXXII. Special Penicillin Issue.

Hemmingson, A.M. 1993. *Q.J. Pharm. Pharmacol.*, 6, 39, 187.

Hewitt, W. 1975. unpublished data.

Hewitt, W. 1977. unpublished data.

Kavanagh, F.W. 1963. In *Analytical Microbiology*, New York: Academic Press.

Kavanagh, F.W. 1968. *Appl. Microbiol.*, 16, 777.

Kavanagh, F.W. 1972. In *Analytical Microbiology*, Vol II, New York: Academic Press.

Pope, C.G. 1945. unpublished information.

5 What Do We Want of an Assay? How Do We Attain Our Goal?

INTRODUCTION

The determination of sample potency may be required for a variety of reasons. It is important for the analyst to consider the purpose of the analysis so that he or she may then select a procedure that will give results of the required quality by a method that is cost effective. A large variety of experimental designs is available to the analyst; these designs and their capabilities will be described elsewhere in this book. Suffice it to say here that they range from simple tests of low precision to tests of higher precision, having more elaborate checks for validity.

In deciding what sort of design to use, some questions that need to be addressed are:

Should the method be:
Rapid? Highly specific? Accurate? and/or Precise?

A brief discussion of these criteria will illustrate how decisions may be made.

In a clinical situation, where samples of body fluids are to be examined in order to decide on further medication, a rapid method is needed; if the patient is being treated with a combination of antiinfectives, then specific methods are indicated. Accuracy and precision may not be of prime importance, as in the case of much pharmaceutical analysis.

When samples of a pharmaceutical product are being examined by the laboratory of a regulatory authority for conformity with official standards, then rapidity is not so important. Standards for precision will be as described in the national pharmacopoeia. It seems logical that for an initial examination of a sample, an assay of medium precision will suffice (i.e., meeting the minimum pharmacopoeial requirement). In the case of samples that then appear to be borderline or below the pharmacopoeial limit, a repeat test of higher precision would be a wise move before challenging the company that is marketing the product.

If samples are being examined by a manufacturing company for either routine control or stability testing, then a degree of precision greater than the pharmacopoeial minimum will be necessary.

PHARMACOPOEIAL INTENTION

Taking, as an example, the *British Pharmacopoeia (BP)*, it is important to read the fine print. The Introduction and General Notices at the beginning of the *BP* (1993) should be read and understood by any analyst concerned with the quality control/assurance of pharmaceutical substances or dosage forms. The Introduction contains a statement of "The Basis of Pharmacopoeial Requirements." This defines precisely the intention of the pharmacopoeial monographs.

A monograph describes the standards with which any sample should comply when tested at any time during its shelf life when stored in accordance with the conditions specified by the manufacturer. Thus, the pharmacopoeial monograph makes allowance for reasonable deviations from the intended active material content in manufacture and an acceptable level of degradation during storage under proper conditions within the allotted shelf life of the product. Pharmacopoeial standards appear to err on the side of leniency to the manufacturer/supplier. It is probably true to say that the regulatory authority is only likely to challenge the manufacturer/supplier when there is very clear evidence that a product is substandard.

CONTROL OF ANTIBIOTIC BULK MATERIALS

The pharmacopoeial guidance in the Introduction is good common sense. It is clearly important to the manufacturer of a pharmaceutical dosage form to have accurate information on the potency of material to be used as the active ingredient. If the potency is underestimated, the manufacturer will increase the weight used and so increase costs; if the potency is overestimated, there is a danger that too little will be used, possibly leading to a product that is subpotent initially or that becomes subpotent during its declared shelf life.

It is logical, therefore, for the manufacturer to require an accurate and precise assay of input materials. Accuracy is demonstrated by the traditional methods that apply equally to chemical, physicochemical, and microbiological assays. The required degree of precision is attained by adequate replication. For example, if confidence limits of ±4 percent ($P = 0.95$) are found for a single assay consisting of a large plate using a 6×6 Latin square design, then probably four such independent assays would yield a mean potency having limits of about $\pm2\%$ ($P = 0.95$). The effect of replication is discussed in Chapter 11. Latin squares are described in Appendix 2.

CONTROL IN ROUTINE MANUFACTURE

It is important to distinguish between what is required of the finished product and how those requirements are attained. A finished product specification may describe tests or limits with which the product would comply *if tested*. It does not necessarily mean that such tests are to be carried out as a part of the routine quality control program. Consider for example a 100-ml bottle of erythromycin ethyl succinate granules for oral suspension containing a nominal 20×250 mg doses. The formulation might include a 5% overage and the release specification might state that each

100-ml bottle contains erythromycin ethyl succinate equivalent to 5.25 ± 0.30 g erythromycin base. The actual schedule of testing for the release of a batch may be (as in the European Union) defined in the product license and, therefore, be subject to agreement with the licensing authority. Logic dictates that the schedule of testing be a rational means of controlling the actual manufacturing process. It is necessary, therefore, to consider the manufacturing process and how it may best be monitored. Suppose, for example, the granule formulation consists of erythromycin ethyl succinate 6.166 g (equivalent to 5.250 g base), together with excipients, to make a total weight of 30.000 g in each 100-ml bottle.

Essentially, there are two possible routes of manufacture:

1. Granules are manufactured to contain 20.55% of erythromycin ethyl succinate, and the fill weight of granules is 30.000 g, corresponding to 5.250 g of erythromycin base.
2. Bottles pass under two filling heads, one of which dispenses 23.834 g weight of excipients and the other dispenses 6.166 g of erythromycin ethyl succinate.

The two rational procedures for controlling manufacture are:

1. For route 1, ensure that the bulk granules are of the correct and uniform composition before filling. The way to do this is not necessarily by microbiological assay. Composition may be determined by liquid chromatography or any method that satisfies first the manufacturer and then the regulatory authority. Regardless of the assay method, it would be necessary to validate the filling procedure to ensure that unmixing did not occur and that a uniform fill within specified limits was obtained.
2. For route 2, naturally, it should be established that the active ingredient conforms to the company specification, which may be a more exacting standard than that of the pharmacopoeial monograph (as suggested in the Introduction to the *BP*). It then remains to ensure that the weight dispensed into each bottle is within acceptable limits. This would be best achieved by validation of the filling procedure during the development of the product, as well as by tests on the amount dispensed immediately before starting and during the production run itself.

If a confirmatory check on the active ingredient content of one or more filled bottles is required (by the regulatory authority), this may be by microbiological assay, liquid chromatography, or other method. Whatever assay method is chosen, the acceptance limits will have to take into consideration both the assay error and the filling error. Such a control procedure should be thought of as a backup for the more effective controls before and during manufacture.

Whichever production procedure and whatever controls are used, the fact remains that the product, if challenged at any time during its shelf life, must comply with the pharmacopoeial requirements and must be found to conform to these standards by microbiological assay. It follows that the really important role of microbiological assay in the overall control of this product is in determination of the potency of the input active ingredient.

RESEARCH AND DEVELOPMENT

This heading covers a miscellany of applications of microbiological assay, from initial development of a new antibiotic to safety and stability of the bulk material and its dosage forms. The principles involved are illustrated by some examples.

YIELD TESTS

In the production of antibiotics, the development of high-yielding strains is of major economic importance. It is of interest to take a historical perspective of this, hence the following quotation from the AOAC Wiley Award Address, "Theory and Practice of Microbiological Assaying for Antibiotics" by Kavanagh (1989), which refers to the state of the art in the 1950s.

> One very important purpose of the assays was to identify new strains of a culture that were at least 5% more productive than the control. Chances of finding such a strain with the 1953 assay were so small that strains could have been selected for further testing by flipping a coin. This would have been at least as productive and much more cost effective. But, such a practice was not scientific. By working slowly so that I did not disturb anyone, I had made all the improvements possible with the facilities available to me by 1963. I had made an analytical instrument from the photometer; standards and samples could be measured accurately with the same special pipette; each rack of 40 tubes was a complete assay of 15 samples done by one person. Now, a 5% change in potency could be detected. Strain selection and media improvement became productive.

In the 21st century, this statement still has a lesson: There is no room for complacency. Attention must be focused on common-sense, practical aspects of the assay, then on assay capability in terms of experimental design and capability of precision.

Consider a program designed to screen a large number of antibiotic-producing cultures so as to select those that are high yielding. A cost-effective approach might be to carry out an initial screening with an assay of relatively low precision that could produce results economically on a large number of cultures. From the initial test, low-yielding cultures could be eliminated, whereas others, perhaps the upper 30%, would go forward for retest by a more precise assay. The choice of cutoff point would be arbitrary; from a knowledge of the precision of the initial screening, the risk of eliminating a potentially high-yielding culture could be assessed.

FOOD SAFETY

In the development of antibiotic feed supplements for animals destined as human food, regulatory authorities require that animal tissues or products, such as eggs and milk, be shown to contain only traces of antibiotic below some defined level.

The experimental principles are the same whether the method of detection and assay is microbiological or chemical. A group of animals receives medicated food or drinking water for a defined period. Then, after a further period, during which only unmedicated food or water is provided, animals are slaughtered and various tissues are examined for traces of the active substance.

In such a program, it is first necessary to demonstrate that the active substance is properly dispersed in the food or drinking water and that it is of adequate stability in the feed or water over the period during which it will be consumed. The levels of antibiotic in a medicated feed might be in the range 400 to 1000 μg/g. To demonstrate these two points, an assay of moderate precision will suffice, e.g., fiducial limits of 95 to 105% ($P = 0.95$).

To determine or detect antibiotic residues in animal tissues or animal products, the expected levels are substantially lower than in the animal feed or drinking water. For this purpose, it is crucial to demonstrate limit of detection and limit of quantitation. Precision is of less importance, and it is suggested that fiducial limits of 80 to 125% ($P = 0.95$) would suffice. A standard-curve assay method would generally be convenient so as to cover a wide range of sample test-solution potencies. Such an assay design is outlined in Chapter 11, under "An Outline of Some Designs Not Illustrated by Examples." However, the analyst should be aware of the appreciable negative bias in the potency estimate due to curvature of the response line when sample potency is outside the range of the standards.

STABILITY OF BULK ANTIBIOTICS

The first essential is that there be a stable reference standard. In the case of long-established antibiotics, there are the international, regional, or national standards that are of accepted stability. Microbiological assay is only one of many tests to be applied to demonstrate stability. Because the objective is to demonstrate only a small drop in potency in a sample stored under temperate conditions over several years, a high-precision assay is essential. For example, it is suggested that a commercially viable antibiotic should not diminish in potency by more than 2% when stored for 12 months under temperate conditions. To detect and quantify a drop of up to 2% would require a very precise assay. Sufficient replicate assays to achieve a mean potency estimate with fiducial limits of 99.0 to 101.0% ($P = 0.95$) would probably be required to demonstrate such a loss. More confidence would accrue as a test continues over several years by combining the observed potency estimates in the calculation of regression and correlation coefficients, if these suggested a small but steady reduction in potency with time.

The precise and, therefore, expensive microbiological assay would be applied at relatively long intervals, for example, 12 months. Additionally, chemical tests for degradation products could be applied at more frequent intervals. Exactly the same principles would apply to accelerated stability tests, although the time scale would be shorter.

STABILITY OF ANTIBIOTIC DOSAGE FORMS

The principles described in the preceeding section are also applicable to dosage forms because long-term stability is a prerequisite for a commercially viable product. Additionally, in the case of a product such as a dry syrup for reconstitution on dispensing, there would be a requirement to demonstrate adequate stability of the liquid preparation when stored for 2 weeks at 4°C. In such a case the acceptable drop may be 5 or perhaps 10%. For this purpose, assays might be carried out at intervals of 1 or 2 days. Assays of somewhat lower precision would suffice; for example, fiducial limits of the mean estimated potency at each time point could perhaps be 97 to 103% ($P = 0.95$). As in the preceding section, calculation of regression and correlation coefficients would increase confidence in the conclusions drawn from the work.

REFERENCES

Kavanagh, F.W. 1989. Theory and practice of microbiological assaying for antibiotics, *J. Assoc. Off. Anal. Chem.*, 72, 6.
British Pharmacopoeia, 1933. Introduction. The basis of pharmacopoeial requirement, p. xxi.

6 General Practical Aspects of Microbiological Assays

INTRODUCTION

Microbiological assay (MBA) is defined here as the estimation of potency of a growth-promoting (GPS) or growth-inhibiting substance (GIS) by comparing its quantitative effect on the growth of a specific microorganism with that of a reference standard of defined potency. It is common belief that being a *biological assay*, MBA is subject to *biological error* and is, therefore, inherently less reliable than chemical or physicochemical methods. However, according to Coomber et al. (1982), it is possible to produce meaningful results and to achieve a precision similar to that of many chemical analytical methods by means of microbiological assay techniques. This may be achieved on a regular basis by the establishment of rigorous control methods that do not leave anything to chance. Vincent (1989) stresses the importance of staff training and good organization to destroy the myth of biological error. The fact of biological error in *macro*biological assays is, of course, accepted; biological variation is not the source of error in *micro*biological assays.

In the sections that follow, general guidance is given on matters of principle in the conduct of microbiological assays. For detailed guidance on all practical aspects of the assay method, the reader is referred to the original writing of Vincent (1989), which describes procedures in meticulous detail.

INOCULUM

For the agar diffusion assay, the inoculum for one large plate typically consists of a suspension containing 10^8/ml or 10^9/ml of cells. These may be vegetative cells from an overnight culture or spores from a stock suspension that may be used over a period of several months when stored at 4°C.

The exact size of the inoculum in terms of viable cells is not highly critical because, for example, a twofold difference in numbers will not make a very large difference to zone diameter and slope of the log dose-response line. However, there must be standardization to ensure that zones have well-defined edges and that the slope of the log dose-response line is adequate.

In the case of the spore suspension, a suitable inoculum size for a newly prepared batch may be determined by inoculating a suitable assay medium at a range of levels, preparing assay plates, then applying standard antibiotic solutions at concentrations that are normally used for assay purposes. A convenient inoculum level would probably be the lowest that gave well-defined edges. This would be expected to give a better slope than higher concentrations that also gave well-defined edges. Once selected, the same inoculum level may be used throughout the lifetime of the batch of spore suspension.

In the antibiotic turbidimetric assay, the inoculum may be about 10^4 to 10^6 viable cells/tube. Although the absolute size of the inoculum is not critical, it will make a difference to the incubation time needed. It is crucial, however, that the inoculum level be the same in all tubes because, if all other conditions are identical, final concentration of cells should be directly proportional to the initial cell concentration. To ensure uniformity of inoculum, the usual procedure is to inoculate the bulk chilled medium, mix thoroughly, then distribute into tubes to which the test solutions have been added previously. This order of addition facilitates thorough mixing of the two liquids.

By way of contrast, the size of the inoculum in the tube assay for a growth-promoting substance is not so critical. By allowing sufficient incubation time, tubes containing equal concentrations of growth-promoting substance should reach the same final cell concentration despite small differences in the inoculum level. Adding one drop of cell suspension to each tube is often the method of achieving inoculation.

TEST SOLUTIONS

There may be occasions when only an approximate potency estimate is required. However, when an accurate potency estimate is the objective of an assay, then test solutions made from both sample and reference standard must be prepared with all the care that is routine in any chemical or physical assay. Thus, a sufficient quantity of material must be weighed to ensure that errors are minimal. Dilutions must be made using appropriate volumetric glassware or a reliable automatic dilution machine.

WEIGHING

Considering first the reference standard, suppose the error in weighing is up to 0.1 mg. Then, if only 10 mg of material are used, the error in this first step can be as much as 1%. If, however, it were possible to use 50 mg, the error from this source would not exceed 0.2%.

If the reference standard is an in-house working standard, it may be available in sufficient quantity that 50 mg could be used on each occasion that a standard test solution is prepared.

Reference standards, such as those of the *USP* or the *EP* Chemical Reference Substances (EPCRS), although intended as working standards, are expensive, and so cost may dictate that smaller quantities, such as 10 or 20 mg, be used. In such cases, the weighing must be done using a suitably sensitive balance and with due care and attention to detail.

The question of moisture content of the reference standard must be considered. The label of the reference standard may direct that it be used "as is" or "after drying". The possibility of change in moisture content during weighing must be considered and appropriate precautions taken.

Glass weighing bottles lose surface adsorbed moisture on drying at about 105°C. After drying, it is customary to allow them to cool in a desiccator over a desiccant such as anhydrous silica gel. When removed from the desiccator, they will immediately begin to adsorb atmospheric moisture. Naturally, the extent of such adsorption depends on the relative humidity of the atmosphere. In weighing a small quantity of an active material (e.g., 20 mg), the weight of moisture picked up by the weighing bottle might be 0.5 mg, which could result in a significant error in weighing (0.5 mg is 2.5% of 20 mg).

Consideration should, therefore, be given to the possible use of aluminium foil on which to weigh the standard. The potential moisture uptake of dried aluminium foil is negligible when compared with that of a small, dried glass weighing bottle. The foil can be folded to make a packet that protects the substance being weighed from contact with the atmosphere.

WEIGHING OR MEASURING FROM A SAMPLE OF THE UNKNOWN

When the sample is a powder or is crystalline, the principles that apply are exactly as for the reference standard. The material will generally be available in quantities such that 50 mg or more of active substance can be weighed. If the sample is not a powder or crystalline, the procedure to be adopted depends on its physical nature.

1. Dilute aqueous solutions. Such samples may be measured using volumetric bulb pipettes calibrated to deliver (D20C). Strictly, such a pipette is designed to deliver the stated quantity of water at 20°C when the water is allowed to drain for fifteen seconds after the water level has dropped to the tip of the pipette, the pipette being held vertically with its tip in contact with the side of the receiving vessel. Provided that the solution is dilute, has about the same viscosity as water, and does not contain surface active agents, such a pipette will deliver the correct quantity of liquid within the permitted tolerances for the particular grade of glassware.
2. Viscous solutions such as syrups and low-viscosity liquids, such as alcoholic solutions. For such solutions, a pipette calibrated to contain (C20C) may be used. These pipettes must be washed out with the diluent so as to remove the entire quantity of sample.
3. Liquid suspensions. These may be treated as in 2 preceeding. It is, of course, necessary to ensure homogeneity by stirring or gentle shaking. The inclusion of air bubbles in the suspension should be avoided.
4. Semisolids containing active material in suspension. This group includes tubes of ointments and creams. Separation may have occurred in the tube;

therefore, the entire contents of the tube should be emptied into a suitable beaker and stirred thoroughly with a glass rod before weighing a portion.

5. Solid dosage forms. Follow the advice given in pharmacopoeias: for example, for tablets, weigh 20 tablets, grind thoroughly using a pestle and mortar, and then weigh accurately a portion of the resulting powder containing approximately the required quantity of active material.

6. Granules for oral suspensions (of antibiotics). In a manufacturing control laboratory, the procedure to be adopted depends on the method of manufacture. This is discussed in the "Control in Routine Manufacture" section of Chapter 5.

In contrast, in the laboratory of a regulatory authority, it is required to determine the content of active material found in the individual bottle tested; thus the whole content of a bottle should be used in preparation of the initial solution. It would be good practice to also record the weight of powder in the bottle used in the test so that the percent w/w could be calculated. Thus, in case of dispute with the manufacturer, it would be possible to distinguish between an incorrect fill weight of a bulk material within its specification and a correct fill weight of a bulk material outside its specification.

DILUTION OF PRIMARY SOLUTION TO TEST SOLUTION LEVEL

When the objective is an accurate potency estimate, dilutions to final test levels must always be made in such a manner that the size of potential errors is minimized and is predictable. If an automatic dilution system is used, it must be validated and shown to be of adequate accuracy and precision.

When dilution is done manually, volumetric glassware must be used, that is, bulb pipettes calibrated to deliver (C2OD) and volumetric flasks.

These may be class A or class B, according to the required accuracy. The permitted tolerances are illustrated by some values presented in Table 6.1 and Table 6.2 for pipettes and flasks respectively. (This information is reproduced with permission from the British Standards Institution. Fuller information may be obtained from the Institution's publications).

It will be noted that for both pipettes and flasks of both classes A and B, the lower the volume, the wider the tolerance. For this reason, observance of the following rule-of-thumb is recommended: Whenever possible, if accurate results are required, do not use bulb pipettes of less than 5 ml capacity and do not use volumetric flasks of less than 50 ml. This may be extended to: Never use a 5-ml pipette if a 10-ml pipette can be used; never use a 50-ml flask if a 100-ml flask can be used, and so on. In practice, the cost of reagents may in some cases be a factor to be taken into consideration so that the selected dilution routine may be a compromise. It is, of course, essential to use properly cleaned glassware; greasy surfaces affect the smooth drainage of pipettes and distort the meniscus of the liquid. Pipettes with chipped tips must be discarded.

Referring again to Table 6.1 and Table 6.2, it will also be noted that the tolerances for class B are generally twice those for the corresponding class A item.

TABLE 6.1
Permitted Variation in Capacity of Volumetric Bulb Pipettes[a]

Capacity	Class A Tolerances		Class B Tolerances	
ml	± ml	± %	± ml	± %
1	0.007	0.70	0.150	1.50
2	0.010	0.50	0.020	1.00
5	0.015	0.30	0.030	0.60
10	0.020	0.20	0.040	0.40
25	0.030	0.12	0.060	0.24

[a] For work of quality, it does not suffice to use straight, graduated pipettes, whether of glass or plastic construction.

Source: Extract adapted from BS1583:1986. Reproduced with permission of BSI under licence number 2001SK/0183. Complete standards can be obtained from BSI Customer Services (Tel. +44 (0) 8996 9001).

TABLE 6.2
Permitted Variation in Capacity of Volumetric Flasks

Capacity	Class A Tolerances		Class B Tolerances	
ml	± ml	± %	± ml	± %
5	0.02	0.40	0.04	0.80
10	0.02	0.20	0.04	0.40
25	0.03	0.12	0.06	0.24
50	0.05	0.10	0.10	0.20
100	0.08	0.08	0.15	0.15
250	0.15	0.06	0.30	0.12
500	0.25	0.05	0.50	0.10

Source: Extract adapted from BS1792:1982. Reproduced with permission of BSI under licence number 2001SK/0183. Complete standards can be obtained from BSI Customer Services (Tel. +44 (0) 8996 9001).

PROBLEMS WITH VERY DILUTE SOLUTIONS

If solutions are very dilute, there may be problems arising from adsorption of active material onto the glass to such an extent that its concentration in the solution is seriously depleted. Kavanagh (1982) drew attention to this problem, citing an example of a 2-μg/ml solution of tylosin that was depleted to the extent of 23% through adsorption onto the surface of a soft-glass bottle. The extent of adsorption depends on the nature of the surface, the nature of the substance, its concentration, and pH of the solution.

THE ASSAY MEDIUM

Formulas for assay media are available from pharmacopoeias and various other publications. Those for the assay of growth-inhibiting substances tend to be rather simple and based on natural nutrients. In contrast, those for the assay of growth-promoting substances are more complex in that they are designed to be free from just one substance needed for growth of the organism — the substance to be assayed. They are therefore synthetic media.

All media are liable to some degradation during heat treatment, which could affect their nutrient qualities and, in the case of solid media, their efficacy as a medium for the diffusion of the active substance in an assay. Media for the assay of both growth-inhibiting and growth-producing substances are available commercially as powders for dissolving in water, then sterilizing.

Consider first media for the assay of growth-inhibiting substances. Because they include natural nutrients, such as bacteriological peptone, beef extract, and yeast extract, their nutrient properties are liable to vary from batch to batch. In solid media, the characteristics of the agar, too, may vary between manufacturers and between batches from the same manufacturer. Each batch of freshly-prepared medium needs to be tested before use. For example, in a batch of agar medium for a growth-inhibiting substances assay, the batch should be tested to see whether it gives clearly defined zone edges and whether the slope b is satisfactory.

This testing may be done conveniently by preparing the new medium and packing it into bottles while stocks of the previous batch remain. Then, in the course of routine testing, run an additional assay with the new batch of medium alongside one using the current batch. Note that the testing of the new medium entails the use of a bottle of medium that has been allowed to cool and set and is then reheated to melt before use. Thus, the bottle from the new batch will have been subjected to the same overall pattern of heat treatment as those bottles that will be used subsequently.

In the case of antibiotics that are a mixture of related active substances, a little-known potential problem may arise due to the composition of the medium. If the active components of sample (unknown) and reference standard were *qualitatively* identical, then the nature of the assay medium would not be a factor that could cause bias in the potency estimate. However, this is not always the case. Many antibiotics have two or more active components; these components may respond differently to the medium. The nature of the agar may be an important feature of the medium.

The agar in the medium may have an important effect on the diffusion of the active ingredients. Different related active components may diffuse at different rates through the agar. If the proportion of the different active components in standard and unknown are not identical, then the resulting relative sizes of their inhibition zones will be dependent on both the concentration of each active component and the influence of the agar. If the medium were made with a different concentration

TABLE 6.3
Relative Potencies of Two Major Components of Polymyxin B1 as Determined Using Agar from Different Sources

Agar Source	Antibiotic Component	Relative Potency
Eiken	Polymyxin B1	100 ± 4.9%
	Polymyxin B2	98 ± 6.1%
Difco	Polymyxin B1	100 ± 5.0%
	Polymyxin B2	129 ± 6.0%

of agar, or with agar from a different source that was qualitatively different, then relative zone sizes for standard and unknown could vary according to composition of the medium.

Such a case was reported by Fujihara et al. (1994). They observed that in the assay of polymyxin B1 there were significant differences in the assay of one sample according to the brand of agar used. Two major components, polymyxin B1 and polymyxin B2, were isolated and their relative potencies determined using assay medium based on (1) Eiken brand agar and (2) Difco brand agar. Setting the potency of polymyxin B1 arbitrarily at 100 percent, the relative potencies found were as shown in Table 6.3.

The question arises, "Which result is correct?" The same problem occurs in any biological assay when one tries to compare unlike materials. It is just not possible to say that one result is correct and the other is incorrect. If disputes arise between laboratories because of such circumstances, the laboratories may be able to agree to assay methods in which every aspect is identical in the two laboratories and thus arrive at compatible conclusions. However, this would not mean that they had arrived at the best result. In such circumstances, it may be advantageous to consider replacing the agar diffusion assay by a turbidimetric assay. A turbidimetric assay may resolve the interlaboratory conflict and provide a more meaningful result.

Problems have been reported in discrepancies between potency estimates by different laboratories assaying erythromycin (Hewitt 1996). There was a fairly constant discrepancy of 6% between the two laboratories. This remained despite critical planning of a collaborative assay in which the same reference standards and the same unknown were used. Similar problems arose between different laboratories assaying polymyxin (Hewitt and Nelis, 1996). Although no work was done to positively identify the cause, differences in the agar must be a possibility.

SELECTION OF LATIN SQUARES AND THE PLATING ROUTINE

A Latin square design is an arrangement of the numbers 1 to n in n rows and n columns in such a way that each number appears once, and once only, in each row and each column. This is a simplified description. The theory of Latin squares is discussed by Fisher and Yates (1938), who give examples of basic designs. In microbiological assays, n commonly has the value of either 6 or 8, giving 6×6 and 8×8 Latin squares, respectively. Variations on the basic designs by randomization lead to vast numbers of possible 6×6 and 8×8 designs. Randomized Latin squares are used for the following reasons:

1. Despite care in the practical aspects of the assay, there may remain some differences throughout the length and breadth of a large assay plate that could influence the zone size; these include variations in thickness of the agar medium, uneven dispersion of the test organism, temperature variations during diffusion, and incubation. The Latin square design tends to compensate for such differences.

2. The differences just described in (1) could tend to be the same in all plates. Thus, in replicate assays of the same unknown, use of different Latin square designs is a further safeguard against bias.

3. Test solutions cannot all be applied to the plate at the same instant. Differences in the time of application necessarily mean differences in length of incubation time. When solutions are applied at a regular rate and in a regular routine — such as starting in row 1 and plating from left to right, then proceeding to row 2 and continuing from left to right, and so on — there will be perfect compensation for time differences. Thus, in an 8×8 assay, the average time of diffusion for all eight positions of test solution number 1 will be the same as that for test solutions 2 to 8.

4. A hazard in manual measurement of zones is *operator anticipation of zone size*. If the operator knows that the zone he or she is reading corresponds to a known test-solution number, there is the danger that, unwittingly, the zone size may be recorded as close to the previous measurement for the same test solution. This is a real danger and may result in sets of observations that are too good to be true, leading to overoptimistic estimates of confidence limits. This danger is obviated by the use of many different designs so that the operator records zone diameter according to position on the plate without knowing at that time the test solution number corresponding to the zone. Any individual laboratory should have available a good number of designs of each size used. Designs should be numbered and selected for use by a truly random process such as a simple computer program to generate random numbers. Vincent (2001) had a choice of 300 8×8 designs in his laboratory.

ASEPTIC TECHNIQUES

Whether or not aseptic techniques are needed must be considered for the various unit operations of an assay.

THE TEST ORGANISM

All operations concerning the maintenance of the master culture must be carried out with precautions to avoid contamination. If the culture were to become contaminated by a faster-growing organism, the contaminating organisms could soon outnumber the original. This could have disastrous results:

1. The contaminating organism might be sensitive to the active substance (growth-inhibiting or growth-promoting) but lack required specificity.
2. The contaminating organism may be insensitive to a growth-inhibiting substance, resulting in uninhibited growth in both turbidimetric and agar diffusion assays.

INOCULATING THE MEDIUM

Precautions to avoid contamination of the assay medium are necessary as are precautions to avoid contamination of a bulk spore suspension that will be used for future assays. However, it is not normally necessary to work in a laminar flow cabinet for such work.

ASSAY PLATES (AGAR DIFFUSION ASSAY)

Glass petri dishes may be sterilized by heating in an oven at 150°C for 1 hour. Large plates generally consist of a plate-glass base with a detachable aluminium frame (held in position by clips) to contain the agar gel and a heavy gauge aluminium lid. Plates, lids, and frames may be cleaned by immersion in a disinfectant solution such as a 2% aqueous solution of Hycolin (a broad-spectrum disinfectant; active ingredients are synthetic phenol derivatives) for 30 minutes. The plates, frames, lids, and clips may then be washed with hot soapy water, followed by rinsing thoroughly with tap water and finally with distilled water, after which they are allowed to drain dry (Vincent, 1992).

At regular intervals, such plates can be heat-treated by stacking them loosely assembled (without clips) in a cold oven and gradually bringing the temperature up to 160°C, then switching off the oven and allowing it and its contents to cool naturally. It is necessary to avoid too-rapid changes in temperature, which could lead to breakages.

ASSAY TUBES (TURBIDIMETRIC ASSAY)

Whether for the assay of growth-promoting or growth-inhibiting substances, closed tubes should be sterilized by dry heat at 150 to 160°C for 1 hour. A convenient closure is a loose-fitting aluminium cap.

DILUENTS

Buffer solutions are normally prepared in large volumes (e.g., 10 liters or more). Freshly distilled or freshly deionized water should be used for their preparation. Even though freshly prepared, such purified water may contain substantial numbers of bacteria, which can multiply rapidly if kept at room temperature. Such nonsterile buffer solutions should be used on the day of preparation. Although they could be stored in a refrigerator for up to 1 week, lack of adequate storage space will normally preclude this option. They may be sterilized in containers of, say, 2-liters capacity; in that case they would need to be labeled with dates of preparation and expiration.

THE SAMPLE

The sample may be, but often is not, sterile. If the sample is a pharmaceutical dosage form that has been manufactured under hygienic conditions, any contamination with microorganisms is likely to be slight. Moreover, being a pharmaceutical, it will be of high potency and require much dilution to produce the test solutions. Thus, dilution with a clean diluent will lead to test solutions that are only very lightly contaminated. Samples other than pharmaceutical dosage forms may be more liable to contamination; if not very potent, they would require less dilution, and thus there could be appreciable contamination of the final test solutions.

TEST SOLUTIONS AND THE EFFECT OF CONTAMINATION

Test solutions are not prepared using aseptic techniques; volumetric glassware and diluents are not normally sterile. All operations should be conducted under hygienic conditions to minimize the level of contamination in the final dilution to test solution level. In the case of the agar diffusion assay, there is no intimate mixing of the possibly contaminated test solution with the inoculated assay medium. To have any adverse effect, the contaminant would have to encroach on the already massively inoculated medium. However, if the test solution were heavily contaminated with a fast-growing organism, the assay could fail.

In the case of either growth-inhibiting substance or growth-promoting substance tube assays, there is intimate mixing of the test solution with the liquid medium so that contamination of the test solution is potentially a greater problem. Nevertheless, when the test solution is prepared hygienically from a pharmaceutical dosage form, the level of any contamination is likely to be very low compared with the substantial inoculation with the test organism. Unless the contamination was by a relatively fast-growing organism, it would be unlikely to compete with the test organism.

APPLICATION OF TEST SOLUTIONS (AGAR DIFFUSION ASSAYS)

Application of test solution to the wells is a procedure of duration ranging from 2 to 3 minutes for a set of, say, six petri dishes to about 15 minutes for a large (8×8) plate, depending on the routine employed.

Exposure of the plate to atmospheric contamination should be minimized. Generally, in the case of petri dishes, this is not a problem in an open but clean laboratory environment. For large plates, it is good practice to keep the nonworking area of the plate covered with the lid during the plating-out operation.

APPLICATION OF TEST SOLUTIONS (TURBIDIMETRIC ASSAYS)

This is a rapid operation, requiring the cap of the assay tube to be removed for just a few seconds. Normally it is sufficient that this be done in a clean laboratory, free from undue air currents. A laminar flow cabinet is not normally needed for this operation.

MEASURING RESPONSES

The measured response is either the size of a zone of inhibition or exhibition in the agar diffusion assay or a measure of turbidity in a tube assay.

THE AGAR DIFFUSION ASSAY

Traditionally, the diameter of each zone is measured to the nearest 0.1 mm. A convenient way of doing this is to view the plate, illuminated from below, against a black background. A suitable plate-reading box is illustrated in Figure 6.1. Machinists' vernier calipers with needles soldered to the jaws may be used for measuring; the points are used to touch the surface of the agar at opposite edges of the circular zone. It is not essential for the two needle points to be in contact when the jaws are closed, corresponding to a reading of zero, nor is it necessary to make any correction to the observed zone diameters, because a constant error makes no difference whatsoever to the calculated potency estimate and fiducial limits. However, it is important that the needles are rigid so that they cannot become bent during the course of reading a plate.

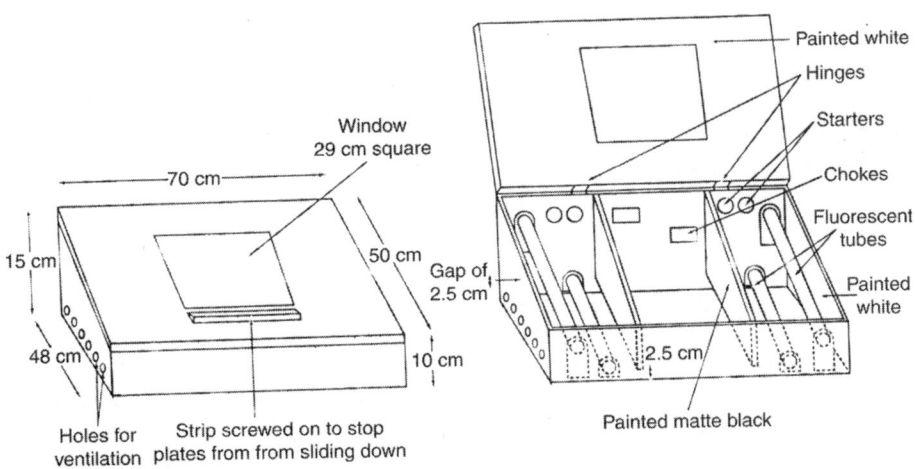

FIGURE 6.1 An illustration of an illuminated plate-reading box. The picture on the left is of the box with the lid closed, as when in use. A large square plate fits neatly in the window. The plate is illuminated from the side and zones are viewed against a matte black background. The picture on the right shows the box with the lid opened, as for maintenance.

As an alternative to hand-held calipers, a magnified image of the zone may be projected onto a screen and the diameter of the image measured.

In measuring zones on a large plate, such as an 8×8 Latin square design, it is customary to start with the zone at the left side of the top row, then read succeeding rows from left to right. Naturally, an objective assessment of the diameter of each zone is necessary, and so it is not permissible to read all eight zones corresponding to a single test solution in succession. To do so would introduce the potential hazard of *operator anticipation of zone size*. Thus, if zones corresponding to a single test solution were read in succession, there might be a temptation to measure the first of the eight and then find (without moving the caliper jaws) that all the other seven were about the same diameter.

Even when plates are read row by row and left to right, there remains the possibility that the operator memorizes a favorite plate design through extended usage, and that he or she, recognizing that the next zone to be read is, for example, high-dose standard, might anticipate the required zone diameter.

An alternative way of measuring zone size without any possible operator bias is by image analysis. The assay plate is placed on an illuminated box and the inhibition zone area is measured by the image analyzer in terms of pixels. A pixel is a unit area of light on a display screen. The magnitude of a pixel is dependent on the scanning system. For purposes of illustration, in one image-analysis system a reading of about 1070 pixels corresponded to a zone diameter of about 22 mm and a reading of about 1560 pixels corresponded to a zone diameter of about 27 mm.

Broadbridge (2001) draws attention to the difficulty in providing uniform illumination over a wide area, such as a 30 cm^2 assay plate. This potential difficulty may be obviated by equipment designed to illuminate an area of the plate, including just one zone at a time, and to move the plate to bring different zones into view in sequence.

The zone area expressed as pixels can be used directly to calculate estimated potency and confidence limits. Alternatively, the software can first convert the area into diameter in millimeters. This step makes the observations directly comparable with the traditional zone diameters but has no other virtue.

The method is highly commended by some users, although it is accepted that it has some limitations. Poor contrast between areas of growth and areas of inhibition is a feature of colistin and polymyxin assays, using *Bordatella bronchiseptica* as the assay organism. In those cases where the method is applicable, very high reproducibility of measurements is claimed by Clontz (2001). In a collaborative study of assays of erythromycin, using *Bacillus pumilis* as the test organism Fraser (1996) achieved confidence limits of ± 2.0 to $\pm 3.0\%$ ($P = 0.95$) for an assay in which one unknown was compared with the standard on a 64-zone large plate. Although this study was not designed to evaluate the zone reading system, it is clear that to achieve such excellent results, the zone sizes must have been read with precision.

THE TURBIDIMETRIC ASSAY

The nature of the response and means of measuring it have been discussed in the "Measurement of Response" section of Chapter 4. Practical aspects of making an accurate measurement of the optical properties of the content of a substantial number of tubes in a realistic period of time still remain to be considered. Measurements of optical properties are commonly made in the actual assay tube, although these are obviously not designed for measurement of transmitted light. Tubes must be as uniform as possible and free from blemishes.

During several hours of incubation, there will have been settling of the cells. The cells need to be redispersed, but without aerating the liquid and producing fine bubbles that would affect the optical properties of the suspension. One suggested procedure is to invert the tubes six times. Another possibility is to stir the content of each tube with a micromagnetic stirring bar. Such bars can be made cheaply and rapidly by taking a piece of steel wire (e.g., from a paper clip) about 8 to 10 mm long and sealing it into a suitable length of a melting-point tube. About 15 seconds of stirring for each tube is probably adequate. The stirring bar may be left in the tube during the measuring operation. In automated systems in which the suspension is withdrawn from the tube and transferred to a cuvette, the transfer system probably provides adequate mixing.

CALCULATION OF POTENCY ESTIMATE AND CONFIDENCE LIMITS

In the testing of a pharmaceutical substance or dosage form for quality control purposes, the records are generally liable to scrutiny by a national regulatory authority, in which case a statistical evaluation of each assay leading to potency estimate and confidence limits will be obligatory. In that case, the only practicable way is to use a computer to do the calculation.

In other circumstances, such as a research or development program, it may not be necessary to do a statistical evaluation. The justification for this statement is that it is not customary (or not possible) in the case of chemical or physicochemical assays to evaluate each individual assay from internal evidence and calculate confidence limits. Why then should individual evaluation be necessary for a well-established and controllable method such as microbiological assay?

Of course, it can be argued that with a computer program, there is no effort in doing the full calculation. The disadvantage is that the program may absolve the analyst from any effort so that he or she may not understand the real meaning of the results of the calculation. It is highly desirable, therefore, that analysts learn how to do the calculation manually and understand the meanings and significance (not the statistical significance) of the various intermediate values that are calculated. Very important variables are the slope, b, and the residual error, s^2. A low value for residual error indicates a precise assay; it also means that variance ratios are liable to be above the arbitrary limits, thus possibly suggesting an invalid assay. It is in such circumstances that the analyst experienced in these calculations is able to make the decision to accept or reject the assay data. This is in accordance with *EP* guidelines (1997, 310) that the analyst makes the decision in the light of the detailed calculations produced by a computer. This topic is raised again in Chapter 9 (when Example 9.1 is calculated according to *EP* guidelines), where it is pointed out that awareness of this principle goes back to at least 1953. Yet commercially available computer software programs exist that still purport to make the decision for the analyst.

A recently developed software package is the Hewitt Bioassays series. This Excel spreadsheet application provides a range of the calculation procedures needed for microbiological assay evaluation. It allows for the computer to make a random selection from a bank of Latin square designs (thus preventing the analyst from using a favorite design that could lead to bias and operator anticipation of zone size). It requires input from the analyst at various stages, such as the preliminary inspection of data and assessment of the tabulated results of the analysis of variance. It thus conforms to the *EP* guidance that the analyst should make decisions based on the detailed calculations produced by a computer. The Hewitt Bioassays series is integrated with a Data Compliance System, the DaCS software of Wimmer Systems, U.S., and so the package complies with the U.S. FDA's electronic records and electronic signature requirements, set forth in Title 21, Part 11 of the Code of Federal Regulations (CFR).

Commercially available software for the calculations should be assessed critically. Consider, for example, the calculation illustrated in Chapter 9 (Example 9.4). Although this is without doubt statistically correct, the tests for differences due to duplicate weighings are not sufficiently sensitive to detect discrepancies that would be unacceptable to a critical analytical chemist.

REFERENCES

Broadbridge, A. 2001. Personal communication.

Clontz, L. 2001. Personal communication.

Coomber, P.A., Jefferies, J.P., and Woodford, J.D. 1982. *Analyst*, 107, 1451.

European Pharmacopoeia. 1997, 5.3, Statistical analysis of results of biological assays and tests, p. 310.

Fraser, N. 1996. Personal communication.

Fujihara, M., Nishiyama, S., and Hasegawa, S. 1994. *Antimicrobial Agents Chemother*, 38:2665.

Hewitt, W. 1996. Unpublished information.

Hewitt, W. and Nelis, H. 1996. Unpublished information.

Kavanagh, F.W. 1982. Personal communication.

Vincent, S. 1989. *Theory and Application of Microbiological Assay* edited by W. Hewitt and S. Vincent, San Diego: Academic Press.

Vincent, S. 1992. Personal communication.

Vincent, S. 2001. Personal communication.

7 Standard Reference Materials

HISTORICAL INTRODUCTION

In the early twenty-first century, we accept the need for reference standards for biological tests without question. It is interesting to look back at the rather slow development of that recognition.

Burn (1950) drew attention to the early definition of the English standard for the *yard*. A yard was defined as "the length of a king's arm." In a museum in Winchester, the ancient capital of England, there are examples of the standard yard. Unfortunately, due to changes of king, these standard yards are not all the same length. This is an admirable demonstration of an attempt to define a standard in purely biological terms.

Despite the inadequacy of the standard yard, in more recent times, investigators have continued to define standards in purely biological terms. For the assay of digitalis, the *frog unit* was devised. The frog unit was that amount of digitalis required to kill 1 g of frog. However, it was found that when a certain preparation of digitalis powder was examined over a period of 10 months, potency ranged in a random way from 1310 to 3300 units per gram. Similarly, a unit for the measurement of potency of insulin was defined as "the amount required to produce convulsions in a fasting rabbit." This unit was abandoned in the year 1925.

In the early days of the development of penicillin, the need for reference standards was perceived. In the agar diffusion assay, it had been observed that zone sizes and the shape of the curve relating dose to zone size were not reproducible from day to day, and so the need to prepare a standard daily curve was apparent. The need for a stable reference standard was clear, and it was suggested that some stable and easily characterized inhibitor, such as mercuric chloride or proflavine, might be used. However, there were arguments against such a course: (1) different strains of test organism would probably respond differently to the proposed standard and penicillin; and (2) even if a standard test organism were used, there was no reason to suppose that if its sensitivity to penicillin changed, its sensitivity to the standard would change in the same way.

The arguments against a substance other than penicillin prevailed, and a purely arbitrary unit was adopted by Heatley (1944) for use in his own laboratory in Oxford. The unit was defined as "that amount of penicillin contained in 1 ml of a certain phosphate buffer solution containing ether." It is not made clear how stable this solution was, but in 1941, a dry sodium salt was standardized against it and was assigned a potency of 42 U/mg. This dry sodium salt became the new primary standard. Later, a barium salt having a potency of 4.4 U/mg was established also as a primary standard. Heatley (1944) noted that these two primary standards showed no detectable loss of potency over many months, even when portions were transported to hot climates. The *Oxford Unit*, as it became known, was adopted by other workers.

It is interesting to note that in these early days, the use of a reference standard was not axiomatic. In a review of assay methods, Garrod and Heatley (1944) stated:

> The method of choice will depend on circumstances, but unless only a semi-quantitative estimate of potency is required, it is highly desirable that a standard penicillin solution be set up at the same time as the unknown so that the potency of the latter may be expressed in terms of the standard.

This was not the first time that a standard reference material had been established for biological assay. Some five decades earlier, Ehrlich had set up a reference material for the assay of diphtheria antitoxin. This was effectively the international standard by which the potency of diphtheria antitoxin was determined until at least 1914.

OFFICIAL REFERENCE MATERIALS

The ultimate reference materials are those established under the auspices of the World Health Organization (WHO).

International Biological Reference Preparations were being established up to 1986. In some cases, potencies were assigned to these. Since 1986, only International Biological Standards are being established; in all cases, a potency in terms of International Units (IU) is assigned, generally on the basis of an international collaborative study.

An International Unit is defined as "the specific activity contained in a defined weight of the relevant International Biological Standard or International Biological Reference Preparation as determined by the WHO Expert Committee on Biological Standardization."

These substances (IBS or IBRP) are intended to provide a "yardstick" against which national authorities may calibrate their own national standards. Member states of the World Health Assembly have agreed to adopt and use WHO standards. The Assembly has reaffirmed that member states of WHO agree to ensure that units of national standards should be made equivalent to appropriate International Units of International Standards. In fact, there are many national standards not calibrated in terms of International Units. Many countries have established national standards; the European Union has established European Pharmacopoeia Biological Reference Preparations (EPBRP).

INTERNATIONAL STANDARDS FOR ANTIBIOTICS

The Oxford unit for penicillin was the forerunner of the first International Unit for an antibiotic. The history of its development is interesting. The following is a quotation from an article by Raper (1978):

The first penicillin standard was prepared in England (in 1941) and had an established potency of 42 Oxford units per milligram. Eventually it was established that the Oxford unit was equal to the biological activity represented by 0.6 microgram of pure sodium penicillin G (benzylpenicillin). An international conference in London in 1944 established this measurement as the international unit. The first sample of the international standard was prepared from à pool of 3-gram samples supplied by each of the major penicillin producers. These were pooled and crystallized as one lot, and its physical characteristics were determined. On this basis, it was established that 1 milligram of crystalline penicillin G contains the equivalent of 1667 units of penicillin activity. The international standard was prepared at the Northern Regional Research Laboratory in Peoria by Dr. Frank Stodola.

According to the WHO Expert Committee on Biological Standardization 20th Report (1968), the first standard for penicillin was established in 1944 with a potency of 0.0006000 mg. The second standard was established in 1952, and the International Unit was 0.0005988 mg. This twentieth report (1968) goes on to state:

The Committee noted that stocks of the second International Standard for Penicillin (established in 1952) were almost exhausted. Since it is possible to characterize benzylpenicillin preparations adequately by chemical and physical methods and since such preparations are suitable for use as reference material for the control of products containing benzylpenicillin, the Committee decided not to replace this International Standard. The Committee agreed, however, that the International Standard for penicillin should continue to be distributed until stocks were exhausted. The Committee requested the WHO Secretariat to consider the need for a preparation of benzylpenicillin of adequate purity to be made available as a chemical reference substance.

In the report of the next meeting (WHO, 1969), the Committee "noted that stocks of the International Standard for Penicillin were now exhausted. In accordance with the decision taken at its twentieth meeting that this standard would not be replaced, the Committee discontinued the International Standard for Penicillin."

Following penicillin, the search for new antibiotics began. In the 1950s to 1970s, a good number were discovered, were developed, and became commercially available. These generally consisted of mixtures of related substances that could not be characterized adequately by chemical or physical methods. New International Biological Standards and International Biological Reference Preparations followed.

As stated earlier, the intended purpose of international reference standards is to provide a standard against which national authorities may calibrate their own national standards. Accordingly, the quantity of any individual standard is very limited, and only small numbers of ampoules of a standard can be made available to national authorities for calibration purposes.

The number of packages of an international standard that is established is intended to last several years, for example, 5 to 8 years.

THE PREPARATION OF INTERNATIONAL STANDARDS FOR ANTIBIOTICS

In setting up an international standard for the first time, it is necessary to select a batch of material typical of the commercially available material. Thus, when the antibiotic is a mixture of related substances, the proportions of the various components should be the same in the proposed standard as they are in typical commercial material. If different commercial materials are of substantially varying proportions of the related substances, then there will be problems. These will be discussed later in this chapter.

When a suitable material has been selected, the packaged standards are prepared under conditions that ensure a high degree of stability. When the standard is to be supplied as a homogenous powder, it is filled into ampules of neutral glass and dried extensively in vacuum, using phosphous pentoxide as a dessicant. Ampoules are then flushed with dry, oxygen-free nitrogen and sealed at a pressure slightly less than atmospheric.

The quantity filled into an ampoule may range from about 20 to 100 mg, depending on availability. The fact that the antibiotics are so thoroughly dried can lead to the problem of rapid uptake of water on opening the ampoule. Care in weighing to avoid this problem is essential.

Alternatively (and now the preferred procedure), antibiotics may be freeze-dried in the ampoule. The first antibiotic to be prepared in this way was bleomycin; 1-gram aliquots of a solution containing 5 mg of bleomycin were filled with precision into the ampoules. These were freeze-dried and then subjected to further desiccation before sealing. Such ampoules are supplied with a designated number of IU/ampoule; the whole content of the ampoule is used without weighing.

ASSESSMENT OF STABILITY AND ASSIGNING POTENCY

Clearly, when the first standard of a new antibiotic is established, the potency must be assigned arbitrarily. When stocks are dwindling, steps must be taken to establish a replacement standard. Each successive edition of a standard should provide continuity of the specific unit of activity. It should be usable in all possible assay situations. Material must be selected that is qualitatively very similar to its predecessor as determined by chemical/physicochemical examination; when the antibiotic is a mixture of related active substances, these should be present in similar proportions to those of its predecessor.

The potency of the proposed new standard relative to that of its predecessor is established by means of a collaborative assay. Typically, the laboratories of about 8 to 12 national control agencies would be invited to participate in the exercise. Although guidance is given on the general conduct of the assay, the assay method is not specified because diversity is desirable. Where possible, the collaborative assay should include both the agar diffusion and turbidimetric procedures. The raw data from the collaborating laboratories is evaluated statistically. Hopefully, the statistical evaluation will demonstrate that results from different laboratories using different assay methods are mutually consistent and that a weighted mean potency with acceptable confidence limits can be assigned to the replacement reference material.

Stability is also assessed by collaborative assay, in which the potency of ampoules stored at elevated temperatures is compared with that of ampoules stored at –20°C.

PROBLEMS ARISING WHEN COMMERCIALLY AVAILABLE ANTIBIOTICS CONTAIN DIFFERING PROPORTIONS OF RELATED SUBSTANCES

It is not uncommon for an antibiotic to consist of a mixture of related active components or *factors*. In the development of a new antibiotic, it would be logical to separate the factors and determine which have the most desirable characteristics in terms of (1) potency against a range of pathogens and (2) safety. Efforts could then be made to produce a commercial product consisting of only the preferred factor, or having a high percentage of the preferred factor. The means of achieving the desired result would almost certainly include screening many strains of the fermentation organism to find a strain giving a high yield of the preferred factor. It may also be possible to include in the fermentation liquor a precursor of the desired factor. A well-known example of this was the inclusion of phenylacetic acid, which was a precursor of benzylpenicillin, in penicillin fermentations.

However, there are cases where the commercially available product remains a mixture of varying proportions of related substances. Neomycin presents a particularly interesting and difficult problem. It includes neomycin B and neomycin C, which are glycosides of the organic base neamine (originally known as neomycin A). In commercially available neomycin, factor B is the major constituent and has greater antibiotic activity than factor C against a range of organisms. There is sometimes a small proportion of neamine that has little activity. Neomycins B and C are of identical molecular weight and differ only in the configuration of an aminoethyl group and were not easily separable by normal analytical procedures.

Lightbown (1961) referred to the difficulty in obtaining agreement between successive assays in either the same or different laboratories. He cites the reason for this as being the varying effect on the two active components of changes in pH, composition of medium, temperature, and age of the test organism.

Sokolski et al. (1964) discussed the different behavior of components B and C in the agar diffusion assay according to composition of the assay medium and especially its salt content. They showed that by choice of a medium of low ionic content and the right test organism, an assay system could be devised in which components B and C appeared equipotent. However, unless it could be shown that neomycins B and C were equipotent against a range of typical infecting organisms *in vivo*, then such an approach disguises rather than solves the problem.

A study of commercially available neomycins from many countries was made for the World Health Organization (1970). It revealed that the then-current International Reference Preparation was unrepresentative of commercially available material with regard to proportions of components B and C.

Wilson et al. (1973) showed by gas-liquid chromatography that in neomycin products available on the Canadian market, the proportion of neomycin C varied from 2.5 to 31%.

Finally, Lightbown (1961) stated the general case:

For heterogeneous materials controlled biologically and assayed against a heterogeneous standard, it must be recognized that there is no true potency for any particular sample. A sample will have a family of potencies depending on the conditions of the assay. These may be distributed about a mode, but the modal value has no intrinsic superiority over any individual value.

Considering the work reported separately by Lightbown and Sokolski, one is bound to note that their observations are based on the agar diffusion assay. Perhaps the problems would not be so great using the turbidimetric assay, thus eliminating the influence of varying diffusion rates.

The problems have been illustrated using an early example, neomycin. Other antibiotics having this same potential for problems of varying composition include bacitracin, erythromycin, and polymyxin. Reference was made to problems with polymyxin in the section on "The Assay Medium" in Chapter 6.

NATIONAL AND REGIONAL REFERENCE MATERIALS

The effort involved in establishing a standard subsidiary to the international standard is very substantial. It entails the selection of suitable material, its packing in a form that will ensure its stability over many years, then a collaborative assay by several laboratories, followed by the statistical evaluation of the raw data and allocation of a potency in accordance with the findings of the statistical evaluation.

Many countries have established national reference standards. However, the effort required by a small country to establish a national standard corresponding to each international standard would be a daunting task. It is not surprising, therefore, that groups of nations have collaborated in the establishment of regional standards. One of the earliest groups to do this has been the European Union, in which 15 nations share common standards that are certified by the European Pharmacopoeia Commission. The standards are held and distributed by the European Pharmacopoeia Commission Secretariat at Strasbourg, France.

In Asia, regional standards have been established under the auspices of the Association of South East Asian Nations (ASEAN). Member states are Indonesia, Malaysia, the Philippines, Singapore, and Thailand. Standards are available from the individual participating laboratories.

IN-HOUSE STANDARDS

Although working standards may be available from national or regional authorities, it is common practice for laboratories to establish their own working standards. The essential features of a working standard are:

1. It should be qualitatively similar to the official standard of the country and to typical commercial material to be assayed.
2. It must be of adequate intrinsic stability.
3. It must be packaged and stored in such a way as to ensure stability.
4. It must be calibrated with sufficient precision to make it a meaningful standard.

In general, a portion of a good quality commercial pharmaceutical material may be set aside for use as an in-house working standard, then packaged and calibrated. However, the requirement for qualitative similarity is sometimes difficult to attain, hence the use of the word *should* rather than *must* in the case of essential feature (1).

Guidance on intrinsic stability is provided by the WHO publication (1986) *Accelerated Stability Studies of Widely Used Pharmaceutical Substances under Simulated Tropical Conditions*. In this study, pharmaceutical substances were subjected to extreme conditions of heat and humidity. The end result of the study was the compilation of two lists of substances, those liable to degradation and those resistant to degradation. An abstract of these lists is presented in Tables 7.1 and 7.2 in the case of antibiotics and vitamins of the B group.

TABLE 7.1
Antibiotics and Vitamins of the B Group
from the WHO Index of Substances
Resistant to Degradation

Amikacin	Nicotinamide
Cyanocobalamin	Nicotinic acid
Erythromycin	Riboflavine
Erythromycin ethylsuccinate	Rifampicin
Erythromycin etearate	Streptomycin sulphate
Folic acid	

TABLE 7.2
Antibiotics and Vitamins of the B Group from the WHO
Index of Degradable Substances

Amphoteracin B	Dicloxacillin sodium monohydrate
Ampicillin sodium	Doxycycline hyclate
Ampicillin trihydrate	Gentamicin sulphate
Bacitracin	Neomycin sulphate
Bacitracin zinc	Nystatin
Benzathine benzylpenicillin	Oxytetracycline hydrochloride
Benzylpenicillin potassium	Paromomycin sulphate
Benzylpenicillin procaine	Phenoxymethylpenicillin
Benzylpenicillin sodium	Phenoxymethylpenicillin potassium
Carbenicillin sodium	Pyridoxine hydrochloride
Cephalexin	Tetracycline hydrochloride
Chloramphenicol sodium succinate	Thiamine hydrochloride
Chlortetracycline hydrochloride	Thiamine mononitrate
Cloxacillin sodium monohydrate	

Of course, those substances that are liable to degradation are also needed as standards. Subject to appropriate packaging and storage conditions, they too may remain sufficiently unchanged to serve as standards over several years.

When the master standard is a biological standard with potency defined in units per milligram, the working standard must be calibrated by microbiological assay. It should be noted that generally one chemical form of the active substance (e.g., one particular salt) is used as standard for other chemical forms. This is illustrated in Table 7.3.

As already stated, a proposed working standard must be calibrated with sufficient precision to make it a meaningful standard. What constitutes sufficient precision depends on the function of the laboratory and purpose of the test. In the case of a pharmaceutical quality control laboratory for a manufacturing operation, it is necessary to have a high degree of confidence in the assigned potency. If, for example, the assigned potency were 2.0% greater than its true potency, this could pose the danger that products could be released for sale even though they are actually about 2% lower than the release specification limit. It is suggested, therefore, that for pharmaceutical quality-control purposes, the potency of a proposed working standard be estimated by comparison with the master standard with a precision such that the fiducial limits of error are not less than 99 and not greater than 101% of the estimated potency. To achieve such precision in a microbiological assay requires a very substantial effort in several replicate determinations of normal precision or a few replicates of a very high-precision assay, such as the 12×12 Latin square design.

In contrast, if the function of the laboratory is to detect levels of antibiotic residues in animal tissues, precision and accuracy of the result are less critical. An assigned potency within 4% of the true potency might be perfectly adequate.

For some antibiotics and vitamins, there is no biological standard, and in the European Union microbiological assay is not an official method for quality control of such pharmaceutical substances or their dosage forms. Nevertheless, microbiological assay may be used for the determination of these substances in clinical or

---

TABLE 7.3
Antibiotic Biological Reference Substances and the Various Chemical Forms Which May Be Assayed Against Them

Reference Substance	Substance that May Be Assayed	Notes
Bacitracin zinc	Bacitracin	
	Bacitracin zinc	
Erythromycin A base	Erythromycin	Turbidimetric assay
	Erythromycin estolate	After hydrolysis
	Erythromycin ethyl succinate	After hydrolysis
	Erythromycin lactobionate	After hydrolysis
	Erythromycin stearate	No hydrolysis required

research laboratories. In such cases, a working standard may be calibrated by chemical or physicochemical methods.

REFERENCES

Burn, J.H. 1950. *Biological Standardisation*, London: Oxford University Press.
Garrod, L.P. and Heatley, N.G. 1944. Bacteriological methods in connection with penicillin treatment, *Br. J. Sur.*, XXXII, Special Penicillin Issue.
Heatley, N.G. 1944. A method of assay for penicillin, *Biochem. J.*, 38, 61.
Lightbown, J.W. 1961. *Analyst*, 86, 216.
Raper, K. 1978. The penicillin saga remembered, *ASM News*, 44, 645.
Sokolski, W.T., Chidester, C.G., Carpenter, O.S., and Kaneshiro, W.M. 1964. *J. Pharm. Sci.*, 53, 826.
Wilson, W.L., Belec, G., and Hughes, D.W. 1973. *Can. J. Pharm. Sci.*, 8, 48.
WHO. 1968. Expert Committee on Biological Standardization 20th Report, WHO Technical Report Series No. 384.
WHO. 1969. Expert Committee on Biological Standardization 21st Report, WHO Technical Report Series No. 413.
WHO. 1970. WHO Expert Committee on Biological Standardization, WHO/BS/70.1001.
WHO. 1986. *Accelerated Stability Studies of Widely Used Pharmaceutical Substances under Simulated Tropical Conditions*. WHO/PHARM/86.529.
WHO. 1996. *General Guideline for the Establishment, Maintenance, and Distribution of Chemical Reference Substances*. WHO/PHARM/96.590.

8 Preliminary Evaluation of Data

INTRODUCTION

In any assay, whether chemical, physicochemical, or biological, it is good practice to look at the raw data and assess whether at first sight it appears to be of the quality expected and, therefore, suitable for processing to calculate an assay result. Such a preliminary inspection of raw data is particularly useful in microbiological assays where those data often consist of a large number of observations distributed in some form of randomized pattern. The data should be unscrambled, then examined to see whether they appear to conform to the basic assumptions of the particular type of assay.

The assumptions for microbiological assays are as follows:

1. The relationship between response and logarithm of dose may be represented by a straight line over the range of doses used (parallel-line assays), or
2. The relationship between response and dose may be represented by a straight line over the range of doses used (slope ratio assays), and
3. The allocation of treatments is by some random process, for example, use of a randomized Latin square in an agar diffusion assay.
4. The responses to any one treatment are normally distributed.
5. The standard deviations of the responses within each treatment group do not differ significantly from one another.

According the current edition of the *International Pharmacopoeia* (*IP*) (1979), referring only to parallel-line assays, it is not possible to verify criteria (1), (4), and (5) in every experiment. Note: The third edition of the *IP* (1979) refers the reader back to the second edition (1967, 755) for guidance on assay evaluation. When it is known from previous experience of the assay that relationship (1) has applied, then it may be reasonably assumed to apply routinely. When there are three or more dose levels, the relationship can be seen readily from the current assay. Similar principles apply to the slope ratio assay.

With regard to criteria (3) and (4), the *European Pharmacopoeia,* (*EP*) (2001) in contrast to the *IP*, does suggest some possible checks for conformity; thus for criterion (4): When deviations from the normal distribution are *suspected* in a series of small samples, the test of Wilk and Shapiro (1968) *may* be used, and for criterion (5): the homogeneity of the variability within the treatment groups *can* be assessed by using either the test of Bartlett (1937) or that of Cochran (1951).

However, the *European Pharmacopoeia* 2000 (1) goes on to say that criterion (4) is a condition that in practice is almost always fulfilled. Furthermore, minor deviations will in general not introduce serious flaws in the analysis so long as there are several replicates per treatment.

The analyst should be aware that in their guidance on the evaluation of biological assays, pharmacopoeias make no distinction between *macro*biological assays, i.e., tests that are generally on small numbers of animals, and *micro*biological assays that involve vast numbers of microorganisms. The statistical tests of the pharmacopoeias are geared to biological variation between the individual animals of small groups. In microbiological assay, we have a different situation in which biological variation is not the source of random variation. Kavanagh (1979) stated, "The agar diffusion assay is not a biological assay; it is a physicochemical assay in which a microorganism acts as an indicator."

These comments are not intended to imply that the pharmacopoeial guidance is inapplicable, but merely to suggest that because of the relatively high quality of microbiological assays, there will be few occasions on which such tests are needed.

COMMON-SENSE INSPECTION

A preliminary inspection of the unscrambled data will give the analyst a good idea about whether an assay is sound and likely to meet the statistical criteria for validity.

The procedure looks at the individual responses to each treatment and asks the following questions: Is the variation of responses to an individual treatment within a reasonable range, in accordance with the analyst's general experience of the method? Are there apparent outliers? If so, do they suggest a practical error such as use of the wrong test solution?

1. For the agar diffusion assay, compare the treatment totals for different doses of a single preparation. Do these totals appear to fit the assumed mathematical model? That is, are log dose-response lines straight? If not straight, is there slight curvature in the direction expected from theoretical considerations? That is, is the difference between responses to medium and high doses slightly less than that between responses to low and medium doses? Is the extent of curvature in accordance with the analyst's general experience of the method? Repeat these questions for every preparation.

    Compare the treatment totals for the same dose levels of standard and unknown. Are the differences similar for each dose level? That is, do response lines appear to be parallel?

2. For turbidimetric assays of growth-inhibiting substances carry out an inspection similar to that for the agar diffusion assay. Expect some curvature as the log dose-response line is part of a sigmoid curve. Is the curvature in accordance with the analyst's general experience of the assay?

3. For turbidimetric assays of growth-promoting substances, do lines appear to be straight? If not straight, is the curvature in the expected direction in

accordance with general experience of the assay and theoretical considerations, such as are discussed in Chapter 3? Do dose-response lines
appear to coincide at zero dose?

When the potency estimate is calculated arithmetically, curvature of the
dose-response line will lead to a biased potency estimate. If curvature is
very slight, and unknown sample and standard potencies are quite close,
then bias will be slight. If curvature is not slight, and standard and
unknown sample responses are not close, then potency should be estimated graphically.

In cases (1) and (3), it is often helpful to draw a graph.

To an analyst experienced in microbiological assay and its evaluation, such a
preliminary inspection will readily reveal any fault in an assay and may also suggest
its cause. For those who are not yet experienced, application of these simple tests
will ensure that they rapidly acquire the skill of assessment of assays by inspection.

It is a good plan to establish control charts that record the values of critical assay
parameters so that previous experience of any particular assay is readily available
for comparison with current findings. An example of a control chart is given in
Figure 8.1. This chart suggests a slight reduction in the slope b from an average of
about 8.4 mm to an average of about 7.5 mm following the change of assay medium
on 11/30/99 (the only date on which the assay medium was changed).

SPECIFIC TESTS FOR ABNORMALITY

The value of inspection as just described is great. The *EP* (1997, 304) referring to
parallel-line assays, states: "After all the responses in the assay have been obtained
they must be inspected." (Strangely, this excellent advice is not included in recent
editions of the pharmacopoeia). This inspection may be supplemented by some
specific tests described in the subsections that follow.

In the foregoing paragraph, the phrases *must be inspected* and *may be supplemented* are of great significance. It is noted that in an example of an antibiotic assay
using petri dishes, the *EP* (1997, 316) states: "[On inspection] The mean variance
for each treatment group ... did not suggest any dependence of the two statistics.
There was no reason to doubt that the conditions in Section 3.1 [of the Pharmaco-
poeia] had been fulfilled."

No tests for validity were applied. (The conditions of Section 3.1 of the pharmacopoeia are conditions 2, 4, and 5 outline at the beginning of this chapter.)

TEST FOR NORMALITY OF DISTRIBUTION

Deviations from normality are not likely and in any event do not seriously jeopardize
the assay. Nevertheless, the *EP* (2000, 265) suggests that *in case of doubt* the Shapiro-
Wilk test *may* be used to check for deviations from normality.

It is not easy to be confident of detecting deviations from normality of distribution by inspection in the case of small numbers. It seems reasonable to suppose that

CONTROL CHART

A. Fixed Parameters

Antibiotic: streptomycin   Assay procedure   SOP0023

B. Variable Parameters

Date	Operator Plate Preparation	Operator Plate Reading	Medium Batch Number	Spore Susp. Batch	Reference Standard	Error Mean Squares	Slope b	g	M	Estimated Potency	Range of % C.L. $P = 0.95$	W	Comments
08/09/99	EJ	AW	AM40/99/06	99/02	strep018	0.0323	8.02	0.0069	−0.00749	98.3%	7.18%	17919	
20/10/99	CH	CH	AM40/99/06	99/02	strep018	0.0327	8.53	0.0062	−0.00473	98.9%	6.79%	20025	
21/10/99	CH	CH	AM40/99/06	99/02	strep018	0.0352	8.41	0.0069	−0.00356	99.2%	7.15%	18069	
30/11/99	AW	CH	AM40/99/07	99/02	strep018	0.0383	7.47	0.0095	0.02170	105.1%	8.40%	13095	*see below
01/12/99	AW	CH	AM40/99/07	99/02	strep018	0.0323	7.76	0.0074	0.01530	103.6%	7.42%	16768	
09/01/00	CH	EJ	AM40/99/07	99/02	strep018	0.0346	7.45	0.0086	0.00856	102.0%	8.00%	14434	
10/01/00	CH	EJ	AM40/99/07	99/02	strep018	0.0292	7.68	0.0068	0.00637	101.2%	7.13%	18178	

Legend: b = mean slope of the response lines; g = index of significance of the slope b; M = the logarithm of estimated potency, W = statistical weight.

*comment: possible significant change in slope on 30/11/99 due to change of media batch.

FIGURE 8.1 A possible form of CONTROL CHART. The design and conditions of assay are very precisely defined by the SOP number. Thus, all assays are directly comparable.

in a well-conducted *micro*biological assay, responses will be normally distributed unless there are outliers, which should be eliminated from the calculation.

The Shapiro-Wilk test described in the pharmacopoeia is for groups of seven or more. This is, therefore, of limited value in microbiological assay where group size is very commonly six, and so the test is not described here.

TEST FOR HOMOGENEITY OF VARIABILITY WITHIN THE TREATMENT GROUPS

This may be checked by either Bartlett's test or Cochran's test. These appear to be little used in microbiological assay and so are not described here. For further information the reader may consult the *European Pharmacopoeia*.

It is noteworthy that Bartlett's test was criticized by Box (1953) as being too sensitive to nonnormality. He compared it with "putting to sea in a rowing boat to see if conditions were fit for an ocean liner to leave port."

DETECTION OF OUTLIERS

The *USP 24* (2000, 1837) gives specific guidance on this topic. The *EP* does not give such specific guidance but suggests that the variances of responses to individual treatments be calculated. A higher-than-usual value would alert the analyst to a possible outlier.

If there are outliers or missing values, these may be replaced using procedures described in the following section. However, the replacement of outliers is a procedure to be used with caution.

The *United States Pharmacopoeia* offers two alternative procedures for the detection of outliers, which are described here verbatim (*USP 24* 2000, 1837–1838).

1. The first criterion is based on the variation within a single group of supposedly equivalent responses. On the average, it will reject a valid observation once in 25 or once in 50 trials, provided that relatively few, if any, responses within the group are identical. Beginning with the supposedly erratic value or outlier, designate the responses in order of magnitude from y_1 to y_N, where N is the number of observations in the group. Compute the relative gap

$$G_1 = (y_2 - y_1)/(y_N - y_1) \tag{8.1}$$

when $N = 3$ to 7

$$G_2 = (y_3 - y_1)/(y_{N-1} - y_1) \tag{8.2}$$

when $N = 8$ to 13

$$G_3 = (y_3 - y_1)/(y_{N-2} - y_1) \tag{8.3}$$

when $N = 14$ to 24

If G_1, G_2, or G_3 exceeds the critical value in Table 1 [Appendix 3, Part A] for the observed N, there is a statistical basis for omitting the outlier ...

2. The second criterion compares the ranges from a series of $k = 2$ or more groups. Different groups may receive different treatments, but all f responses within each group represent the same treatment. Compute the range from each group by subtracting the smallest response from the largest within each of the k groups. Divide the largest of the k ranges by the sum of all the ranges in the series. Refer this ratio R_* to Table 2 [Appendix 3, Part B]. If k is not larger than 10, use the tabular values in the upper part of Table 2 [Appendix 3, Part B]; if k is larger than 10, multiply R_* by $(k + 2)$ and interpolate, if necessary, between the tabular values in the lower part of Table 2 [Appendix 3, Part B]. If R_* exceeds the tabular or interpolated value, the group with the largest range is suspect and inspection of its components will usually identify the observation, which is then assumed to be aberrant or an outlier. The process may be repeated with the remaining ranges if an outlier is suspected in a second group.

REPLACEMENT OF MISSING VALUES

The question of rejection and replacement of values is a delicate one. It would be wrong to replace a value simply because we did not like it. Conversely, it would be equally wrong to include a value that was clearly erroneous such as might arise by use of the wrong test solution.

When a value is discarded, it must to be replaced by a substitute value so that the calculation may be continued in accordance with its established pattern. The substitute value is calculated in a way that takes into consideration not only the potency of the test solution but any bias that may arise due to its position in the experimental design, for example, different plates in a petri-dish assay or row and column variation in a Latin square design.

Methods for calculation of substitute values are given in the *USP 24* (2000, 1838) and the *EP* (2000, 270). The methods are those of Emmens (1948). Different procedures apply according to the experimental design. They should be used with discretion — a virtue not possessed by computers. This is made clear in the *United States Pharmacopoeia* itself.

The sole purpose of such a substitute value is to facilitate the calculation; it does not add any information to the data available, and so in the statistical evaluation that follows, one degree of freedom is lost for each value substituted.

The lost degree of freedom refers to the residual error and leads to a slightly greater value for error mean squares as well as a slightly greater value for t; these both contribute to a slight widening of the confidence limits.

An illustration of such calculations of replacement values is given in Example 9.5 of Chapter 9, in which one outlier is replaced simply by the mean of the other three responses to the same treatment.

REPLACEMENT OF A MISSING VALUE IN A PETRI DISH ASSAY

For assays consisting of randomized sets, such as a petri dish assay or a Latin square design assay, use the procedure (b) that is described in the *USP 24* (2000). For a petri dish assay, calculate the replacement value y' as:

$$y' = \frac{fT'_r + kT'_t - T'}{(f-1)(k-1)} \tag{8.4}$$

where

f = number of sets (assay plates)
k = number of treatments
T'_r = the incomplete total from the plate having the missing value
T'_t = the incomplete total from the treatment having the missing value
T' = the incomplete totals from the assay as a whole

REPLACEMENT OF A MISSING VALUE IN A LARGE PLATE LATIN SQUARE DESIGN ASSAY

The *United States Pharmacopoeia 24* describes a procedure for application when the single assay consists of more than one plate using a Latin square design.

The replacement value is calculated as

$$y' = \frac{k(n'T'_c + T'_r + T'_t) - (2 \times T')}{(k-1)(n'k-2)} \tag{8.5}$$

where

n' = number of Latin squares with k rows in common
k = number of treatments
T'_c = incomplete total from the column having the missing value
T'_r = incomplete total from the row having the missing value
T'_t = incomplete total from the treatment having the missing value
T' = incomplete total from the assay as a whole

For calculation of a replacement value, the *European Pharmacopoeia* gives the expression

$$y' = \frac{k(B' + C' + T') - (2 \times G')}{(k-1)(k-2)} \tag{8.6}$$

in which B' and C' replace T'_c and T'_r of the *USP* expression and $n' = 1$. That is, the *EP* expression is the specific case of the *USP* expression for just a single assay plate.

SUMMARY AND CONCLUSIONS

Pharmacopoeias give sound advice on the preliminary assessment of the worth of raw data from biological assays in general. The first step is usually the unscrambling of randomized information so that responses to individual treatments are presented in columns so as to permit intracolumn and intercolumn inspection, as well as comparison of column (treatment) totals. An inspection of such data is very helpful in assessing whether the results obtained are compatible with the basic assumptions of the assay that were stated at the beginning of this chapter.

The visual inspection can be supplemented by some specific tests. Useful tests that are very simple to apply with the aid of a computer program include (1) a calculation of the variance of observations for each individual treatment and (2) the two *USP* tests for outliers. The need for caution in replacing outliers is stressed again.

*Micro*biological assay is a special case and is not subject to the large biological errors of *macro*biological assay. The simple inspection procedures outlined in this summary and conclusions should be all that is needed in most cases before proceeding to the analysis of variance and calculation of confidence limits of the potency estimate.

REFERENCES

Bartlett, M.S. 1937. Properties of sufficiency and statistical tests, *Proc. Roy. Soc. London,* Series A 160, 280.

Box, G.E.P. 1953. *Biometrika*, 40, 318.

Cochran, W.G. 1951. Testing a linear relation among variances, *Biometrics*, 7, 17.

Emmens, C.W. 1948. *Principles of Biological Assay*, London: Chapman and Hall.

European Pharmacopoeia 1997 (1) p. 304.

European Pharmacopoeia 1997 (2) p. 316.

European Pharmacopoeia 2001 (1) p. 305.

European Pharmacopoeia 2000 (2) p. 270.

International Pharmacopoeia 1979 Vol. 1, *General methods of analysis*, 3rd ed.

International Pharmacopoeia 1967 2nd ed. (1) p. 755.

Kavanagh, F.W. 1979. Personal communication.

The United States Pharmacopoeia 24 (1) p. 1837.

The United States Pharmacopoeia 24 (2) p. 1838.

Wilk, M.B. and Shapiro, S.S. 1968. The joint assessment of normality of several independent samples, *Technometrics*, 10, 825.

9 Parallel-Line Assays — Some Designs and Their Evaluation

INTRODUCTION

Guidance on assay design and evaluation is given in pharmacopoeias. Reference is made here to the *BP, British Pharmacopoeia*; the *EP, European Pharmacopoeia*; the *IP, International Pharmacopoeia*; and the *USP, United States Pharmacopoeia*. The aim of this chapter is to draw attention to such guidance and to expand on it, showing some fully worked examples.

The first 17 pages of this chapter are an attempt to explain the rather tedious official calculations of the European and United States pharmacopoeias. Personally, I would never follow these exactly; I prefer the pattern of calculations exemplified in the latter part of the chapter. These calculations lead to exactly the same results and, I believe, are more readily comprehensible. However, the official calculations exist and so should be explained. Do not be discouraged. You can skip the detailed calculations of these 17 pages (but not the general comments), then go on and start at the section titled "Back to Basics."

The *BP* gives some general guidance on assay design and criteria for validity but refers the reader to the *EP* for more detailed guidance.

The second and third editions of the *EP* (1993, 1997), gave examples of evaluation of two antibiotic agar diffusion assays:

1. An assay using petri dishes that compares one sample with a standard, both preparations appearing at three dose levels forming a geometrical progression. This example is used here in the next section in this chapter.
2. An assay using a large plate on which two samples are compared with the standard at three dose levels. Test solutions are arranged on the plate in accordance with a 9 × 9 randomized Latin square design. This design appears not to be used very widely and so is not discussed further here. However, its principles are common to other designs illustrated in this text.

These were both replaced in the 2000 Supplement to the *European Pharmacopoeia*, third edition of the *EP*, which gave two antibiotic assays:

1. An agar diffusion assay using a large plate on which one sample is compared with the standard at three dose levels. Test solutions are arranged on the plate in accordance with a 6 × 6 randomized Latin square design. This design is widely used in Europe.
2. A turbidimetric assay with four dose levels and a replication of five (five tubes for each test solution).

The same two examples appear in the 2001 Supplement to the third edition. The current edition (1979) of the *IP* does not include any update on the calculations of biological assay but refers the reader to the second edition (1967). The latter gives a worked example of a three-dose level assay using a 6 × 6 Latin square design.

The *USP 24*, in <111>, Design and Analysis of Biological Assays, gives much valuable guidance on calculation procedures for biological assays in general, but it does not give an actual worked example of a microbiological assay evaluation.

The principles of statistical analysis will be illustrated here using a variety of examples of different assay designs. With the exception of Example 9.1 (introduced in the next section of this chapter), the calculation procedures shown here are not identical with those of any of the pharmacopoeias. They draw on the guidance given by Lees and Tootill (1955a, b, and c) and by Finney (1978), in which an analysis of variance leads to a value for residual error-mean squares, s^2. In the opinion of the writer, these methods are far easier to comprehend than those given in the pharmacopoeias. They enable the analyst to understand the evaluation rather than just follow instructions.

The figure for s^2 is then substituted in one of two expressions:

1. The approximate method — an expression that gives a value for $V(M)$, the variance of the logarithm of the potency estimate. From $V(M)$, it is a short step to obtain the confidence limits of the potency estimate.
2. The exact method — a more complex expression that leads directly to confidence limits.

The approximate method (which is applicable only when g, the index of significance of the slope, is less than 0.1) has been widely used in microbiological assay. Now that computers are readily available to analysts, the additional labor of using the exact method has disappeared. The current edition of the *EP* quotes only the exact method.

Once again, this is a generalization for biological assays. In the special case of *micro*biological assays for which g is almost invariably not only less than 0.1 but less than 0.01, both calculation procedures may be expected to give percentage confidence limits differing only in the second decimal place.

Both methods will be illustrated here in only one example (Example 9.2). Rather than burden the reader with complex calculations, all other examples will use the approximate method, which serves to illustrate principles perfectly and is easier to comprehend. For routine work, however, analysts may follow official guidance and use computer programs incorporating the exact method.

A THREE-DOSE LEVEL ASSAY USING PETRI DISHES (EXAMPLE 9.1)

This example is taken from the *EP* (1997). It is the assay of an antibiotic (unnamed). Test solutions were made from the unknown on the basis of its assigned potency of 1500 U/ml so as to be the same as the nominal potencies of the standard test solutions. Each of the six treatments was applied once in each dish. Zone diameters (in mm × 10) are presented in Table 9.1, together with plate (row) totals, treatment (column) totals, and the ranges, means, and variances of the values for each treatment.

EXAMPLE 9.1   AN ASSAY OF AN ANTIBIOTIC (UNNAMED)

Test organism: not stated

Design: petri-dish assay

Dose ratio: 2:1; nominal standard dose levels: 2, 4, and 8 U/ml.

The preliminary inspection of the data of Table 9.1 that is shown here includes items that we have found useful over many years (marked *) and items suggested by the *EP* (marked **).

TABLE 9.1
The Raw Data from an Antibiotic Assay for Example 9.1 Using Three-Dose Levels of Each of Two Preparations on Each of Six Petri Dishes

Plate	Standards			Unknowns			Block Totals
	S_1	S_2	S_3	U_1	U_2	U_3	
1	176	205	235	174	202	232	1224
2	178	208	238	175	206	234	1239
3	178	207	237	177	203	236	1238
4	175	205	235	173	201	232	1221
5	176	206	235	174	204	231	1226
6	174	204	236	170	202	229	1215
Total	1057	1235	1416	1043	1218	1394	
Mean	176.2	205.8	236.0	173.8	203.0	232.3	
Range	4	4	3	7	5	7	
Variance	2.6	2.2	1.6	5.4	3.2	5.9	

OBSERVATIONS

* Treatment ranges: The lowest and highest ranges for treatments are 3 and 7 (0.3 mm and 0.7 mm), respectively. General experience of the agar diffusion assay suggests that these are satisfactory ranges.
* Comparison of individual slopes for preparations: These are calculated from treatment totals as

$$\text{for standard, } 1416 - 1057 = 359$$

$$\text{for unknown, } 1394 - 1043 = 351$$

Experience indicates that this agreement is very satisfactory: Allowing for acceptable random error, the lines may be regarded as parallel.

* Evidence of curvature:

1. Compare slopes for the two intervals of standard, calculated as

$$\text{high} - \text{medium, } 1416 - 1235 = 181$$

$$\text{medium} - \text{low,} \quad 1235 - 1057 = 178$$

2. Compare slopes for the two intervals of unknown, calculated as

$$\text{high} - \text{medium, } 1394 - 1218 = 176$$

$$\text{medium} - \text{low,} \quad 1218 - 1043 = 175$$

The near identity of slopes over two dose intervals for both standard and unknown shows that this assay conforms to the ideals of the model on which a parallel-line assay is calculated and, therefore, is valid.

** Consider the mean and variance for each treatment group: their values did not suggest any interdependence of the two statistics.

(Note that in this example, the *European Pharmacopoeia* (1997) relied solely on inspection to consider possible interdependence of means and variance; it was not deemed necessary to apply, for example, Bartlett's test.)

From this preliminary inspection, there is no reason to doubt that the conditions described in Chapter 8 have been fulfilled; there are no apparent outliers, and so continuation with the calculation is justified.

EXAMPLE 9.1 CONTINUED: THE CALCULATION ACCORDING TO EUROPEAN PHARMACOPOEIA (EP) GUIDANCE

Procedure for analysis of variance: Referring to the *EP*, Table 3.2.3.- II. — "Formulae for assays with three doses of each preparation" (1997, 306):

Step 1. Obtain the treatment totals S_1, S_2, S_3, U_1, U_2, and U_3 (the column totals from Table 9.1).

Step 2. Obtain linear contrasts for standard and unknown as:

$$S_3 - S_1 = L_S \tag{9.1}$$

and

$$U_3 - U_1 = L_U \tag{9.2}$$

respectively.

Step 3. Obtain quadratic contrasts for standard and unknown as

$$S_1 - 2S_2 + S_3 = Q_S \tag{9.3}$$

and

$$U_1 - 2U_2 + U_3 = Q_U \tag{9.4}$$

respectively.
Now, in accordance with steps 4 to 9 and referring to the *European Pharmacopoeia*, Table 3.2.3.- IV. — "Test of validity" (1997, 307), calculate degrees of freedom f and sums of squares according to source of variation where

number of preparations, $h = 2$
number of treatments, $k = 6$
number of dose levels, $d = 3$
number of replicates for each treatment, $n = 6$
total number of responses in the assay, $N = 36$
total of responses to standard preparation, $S = 3708$
total of responses to unknown preparations, $U = 3655$
and K is the correction term, also known as *mean correction* or *MC*.

Step 4. Calculate the correction term K as

$$K = (\Sigma y)^2 / N \tag{9.5}$$

where Σy, the sum of the treatment totals, is obtained as $S + U$, and N is the total number of responses in the assay.

Step 5. For preparations, the degrees of freedom, $f = h - 1$: sum of squares is

$$[(S^2 + U^2)/3n] - K \tag{9.6}$$

Step 6. For linear regression, the degrees of freedom, $f = 1$: sum of squares is

$$E = (L_S + L_U)^2/2nh \qquad (9.7)$$

Note that the symbol E that was introduced in "Principle of Calculation of Potency Estimate" in Chapter 2, has a quite different meaning.

Step 7. For nonparallelism, the degrees of freedom, $f = h - 1$: sum of squares is

$$\left[\left(L_s^2 + L_u^2\right)/2n\right] - E \qquad (9.8)$$

Step 8. For quadratic regression, the degrees of freedom, $f = 1$: sum of squares is

$$Q = (Q_S + Q_U)^2/6nh \qquad (9.9)$$

where

$$Q_S = 3 \text{ and } Q_U = 1$$

so that

$$Q = (3 + 1)^2/(6 \times 6 \times 2) = 0.222$$

Step 9. For difference in quadratics, the degrees of freedom, $f = h - 1$: sum of squares is

$$\left[\left(Q_s^2 + Q_u^2\right)/6n\right] - Q \qquad (9.10)$$

so that

$$[(3^2 + 1^2)/(6 \times 6)] - 0.222 = 0.056$$

Now referring to the *European Pharmacopoeia*, Table 3.2.3.-V. — "Estimation of residual error" (1997, 308), estimate the residual error by steps 10 – 13

Step 10. For treatments, the degrees of freedom, $f = k - 1$: sum of squares is

$$\left[\left(S_1^2 + S_2^2 + ...U_n^2\right)/n\right] - K \qquad (9.11)$$

Step 11. For blocks (plates), the degrees of freedom, $f = n - 1$: sum of squares is

$$\left[\left(R_1^2 + R_2^2 + \ldots R_n^2\right)/n\right] - K \qquad (9.12)$$

where each R is the total of zone diameters of a single plate.

Step 12. For total deviations, the degrees of freedom, $f = N - 1$: sum of squares is

$$\Sigma y^2 - K \qquad (9.13)$$

Step 13. For residual error, the degrees of freedom, f, and sum of squares are both obtained by difference as

$$\text{residual error} = \text{total} - \text{treatments} - \text{blocks} \qquad (9.14)$$

From the data in Table 9.1, carry out the calculations of steps 1 to 13 using Equations (9.1) to (9.14)

Step 1. Treatment totals from Table 9.1 are

$$S_1 = 1057, S_2 = 1235, S_3 = 1416$$
$$U_1 = 1043, U_2 = 1218, U_3 = 1394$$

Step 2. Obtain linear contrasts for standard and unknown by Equations (9.1) and (9.2), respectively, as

$$L_S = 1416 - 1057 = 359$$
$$L_U = 1394 - 1043 = 351$$

Step 3. Obtain quadratic contrasts for standard and unknown by Equations (9.3) and (9.4), respectively, as

$$Q_S = 1057 - (2 \times 1235) + 1416 = +3$$

and

$$Q_U = 1043 - (2 \times 1218) + 1394 = +1$$

Step 4. Obtain the correction term K by Equation (9.5) as

$$K = (\Sigma y)^2/N = 7363^2/36 = 1,505,938.028$$

where

Σy, the sum of the treatment totals $= S + U = 3708 + 3655 = 7363$

Step 5. The sum of squares for preparations is obtained by Equation (9.6) as

$[(1057 + 1235 + 1416)^2 + (1043 + 1218 + 1394)^2]/(3 \times 6) - 1,505,938.028$
$= 1,506,016.056 - 1,505,938.028 = 78.028$

The corresponding value of f is $2 - 1 = 1$.

Step 6. For linear regression, obtain E by Equation (9.7) as

$$E = (359 + 351)^2/(2 \times 6 \times 2) = 21,004.167$$

The corresponding value of f is 1.

Step 7. Obtain sum of squares for nonparallelism by Equation (9.8) as

$(359^2 + 351^2)/(2 \times 6) - 21,004.167 = 21,006.833 - 21,004.167 = 2.666$

The corresponding value of f is $2 - 1 = 1$.

Step 8. For quadratic regression, obtain Q by Equation (9.9) as

$$Q = (3 + 1)^2/(6 \times 6 \times 2) = 0.222$$

where

$$Q_S = 3 \text{ and } Q_U = 1$$

The corresponding value of f is 1.

Step 9. Obtain sum of squares for difference in quadratics by Equation (9.10) as

$$[(3^2 + 1^2)/(6 \times 6)] - 0.222 = 0.056$$

The corresponding value of f is $2 - 1 = 1$.

Step 10. Obtain sum of squares for treatments by Equation (9.11) as

$$\left[\left(1057^2 + 1235^2 + 1416^2 + 1043^2 + 1218^2 + 1394^2\right)\right]/6 - (1,505,938.028)$$

$$= 1,527,023.167 - 1,505,938.028 = 21,085.139$$

The corresponding value of f is $6 - 1 = 5$.

Step 11. Obtain sum of squares for blocks (plates) by Equation (9.12) as

$$\left[\left(1224^2 + 1239^2 + 1238^2 + 1221^2 + 1226^2 + 1215^2\right)\big/6\right] - 1,505,938.028$$

$$= 1,506,013.833 - 1,505,938.028 = 75.805$$

The corresponding value of f is $6 - 1 = 5$.

Step 12. Obtain sum of squares for total deviation by Equation (9.13) as

$$= \left(176^2 + 205^2 + 235^2 + \ldots + 229^2\right) - 1,505,938.028$$

$$= 1,527,127 - 1,505,938.028 = 21,188.972$$

The corresponding value of f is $36 - 1 = 35$.

Step 13. Obtain residual error by Equation (9.14) as

$$21,188.972 - 21,085.139 - 75.805 = 28.028$$

The corresponding value of f is $35 - 5 - 5 = 25$.

The values obtained in the foregoing calculation for degrees of freedom and deviation squares are now summarized in columns 2 and 3 of Table 9.2. Mean squares (where appropriate) and variance ratios are also tabulated. The variance ratio is the ratio of the mean squares for an individual source of variation to the mean squares for the residual error.

Discussion of Assay Validity Based on the Analysis of Variance

First, a comment on the interpretation of variance ratios in general is appropriate. Pharmacopoeial requirements for validity of parallel-line assays concern variance ratios for only linear regression (which should be large) and for nonparallelism and nonlinearity (which should be small). Variance ratios are obtained for other features such as preparations, plates (in the case of small plate assays), and rows and columns (in the case of large plate assays using Latin square designs). There are no official requirements concerning these variance ratios; their values are informative and may provide warnings that assay conditions could be improved.

The importance of the analyst's interpretation of the variance ratios cannot be overemphasized. Whereas the *European Pharmacopoeia* in the last years of the twentieth century drew attention to this need, its existence was recognized almost 50 years earlier.

TABLE 9.2
Summary of the Analysis of Variance for Example 9.1 as Calculated by the _European Pharmacopoeia_ Method (EP). A Three-Dose Level Assay Using Six Petri Dishes

Source of Variance	d.f.	Sum of Squares	Mean Squares	Variance Ratio	Limiting Value	At Arbitrary Probability Level, P	Calculated Probability
Preparations	1	78.03	78.03	69.61	<4.24	0.050	<0.0001
Regression	1	21,004.17	21,004.17	18,737.00	>13.88	0.001	<0.0001
Parallelism (deviations from)	1	2.67	2.67	2.38	<4.24	0.050	0.1355
Quadratic curvature	1	0.22	0.22	0.20	<4.24	0.050	0.6586
Difference of quadratics	1	0.06	0.06	0.05	<4.24	0.050	0.8249
Subtotal	5	21,085.15					
Treatments	5	21,085.14					
Plates	5	75.81	15.16	13.54	<2.60	0.050	<0.0001
Residual error	25	28.02	1.12	1.00			
Total	35	21,188.97					

Note: Limiting values in column 6 refer to the corresponding arbitrary probability level of column 7.

The following quotation is from Finney (1978):

It is important to avoid any automatic rule of rejecting assays on account of nonlinearity or other aspects of statistical invalidity. As Humphrey et al. (1953) have emphasised, a rule based solely on individual significance tests would merely result in the most precise assays being rejected! A truly linear regression is a rarity, and to penalise all assays in which high precision detects nonlinearity is a folly. To formulate an ideal policy is difficult.

Pharmacopoeial interpretation of variance ratios (or F values) has, traditionally, been on the basis of comparison of the F value with an arbitrary limit obtained from statistical F tables. This is exemplified by the official requirements applicable to the example under consideration now:

1. Linear regression must be highly significant, that is, the probability must be less than 0.01. This is ascertained by comparing the variance ratio found (18,737.0) with the value in the 1% points table for $n_1 = 1$ and $n_2 = 25$ where n_1 and n_2 are the number of degrees of freedom for regression and for residual error, respectively. The tabulated value is 7.77, so the value found vastly exceeds the arbitrary limit showing that regression is highly significant.

2. Nonparallelism and nonlinearity must be nonsignificant, that is, the probability must be not less than 0.05. Taking the case of parallelism, the variance ratio found was 2.38; this was compared with the value in the 5% points table for $n_1 = 1$ and $n_2 = 25$. The tabulated value is 4.24, so the value found is below the arbitrary limit, and there is no significant deviation from parallelism.

Now returning to consideration of regression, in practice, in microbiological assays the variance ratio for linear regression is invariably highly significant. In Table 9.2, it is compared with the much more stringent criteria of the 0.001 level. The limiting value for regression, 13.88, was obtained from the variance ratio table for the 0.1 percent level ($P < 0.001$) for $n_1 = 1$ and $n_2 = 25$. The value 18,737.0 greatly exceeds this limit, and so it is very satisfactory.

A microbiological assay would never become established if there were going to be any doubt about the significance of regression, and so this is a criterion the analyst is not really going to need to worry about.

Apart from this test, the significance of regression is checked later by calculation of the statistic g or C — integral parts of the overall calculation. Regarding deviations from linearity, examples in the second and third editions of the *European Pharmacopoeia* (1993 and 1997) showed variance ratios for "quadratic curvature" and for "difference of quadratics," each with one degree of freedom, as is done here. In the year 2000 supplement to the third edition, these two sources of variation are combined as "nonlinearity" with two degrees of freedom. The reason for this change is not clear; it seems doubtful that there could be strong reasons for preferring either of the two alternatives. Here, the individual values for variance ratio at 0.20 and 0.05, respectively, are well below the limiting value of 4.24. If the values were pooled, the mean squares would be $0.28/2 = 0.14$. This would be below the new limiting value of 2.38 for $n_1 = 2$ and $n_2 = 25$. Thus, the same conclusion would be reached in this case. Still to be considered are the implications of the values for preparations and plates; since there are no pharmacopoeial limits, these will be compared with warning limits.

At the arbitrary 5% level ($P < 0.050$), the warning value for preparations is 4.24, obtained in the same way as for the limiting value for parallelism. The variance ratio found, 69.61, greatly exceeds the warning limit. This tells us that there is a highly significant difference between the potencies of the two preparations. It merely confirms what we already know — the unknown sample has a potency estimated to be 1400 U/ml, which is 93.3% of that of the reference standard. This does not indicate a great problem. Indeed, there is only a very small problem in the case of this assay. The significant difference in potency has two effects:

1. The confidence limits will be very slightly wider than if the relative potencies had been 100.0%.
2. As there is some curvature of the response line, there will be some bias to the potency estimate. However, this is a symmetrical assay, and it has been demonstrated that such bias is likely to be very slight in symmetrical

assays when potency of the unknown deviates from that of the standard by ±20% or even more; see the "Bias Due to Curvature" section in Chapter 11. It will be seen in subsequent examples that this warning of significant differences between preparations is often ignored. One is bound to wonder whether the test serves any useful purpose.

The warning value for plates, 2.60, was obtained from the variance ratio table for the 5% ($P < 0.050$) level for $n_1 = 5$ and $n_2 = 25$. Although the ratio 13.54 is much greater than the warning value, this does not invalidate the assay because each plate includes each treatment once and once only, so each plate may be regarded as an independent assay. The significant variance ratio indicates some real difference between assay plates — perhaps a difference in diffusion and/or incubation temperatures or difference in the thickness of the agar layer.

On the basis of these tests, there is no cause to doubt the validity of the assay, and so we may proceed to calculate the width of the confidence limits of the potency estimate.

As stated earlier, interpretation of F values has traditionally been on the basis of comparison with an arbitrary limit obtained from statistical F tables. Although the *European Pharmacopoeia* (2000) includes a simple F-distribution table, it recommends the use of statistical functions included in computer programs to calculate the actual probability from the F ratio found and the two numbers of degrees of freedom. The pharmacopoeia also provides a generating procedure for the calculation of probabilities.

In all the summaries of analysis of variance in this book, both the limiting values for F and the calculated probabilities are shown. Probabilities were calculated using the FDIST program of Excel.

Calculation of Potency Estimate (*EP* 1997, 317)
European Pharmacopoeia

Step 1. Contrast the unknown with the standard:

$$\bar{y}_u - \bar{y}_s = \frac{U - S}{3n} = \frac{3655 - 3708}{3 \times 6} = \frac{-53}{18} = -2.944444$$

Step 2. Calculate the slope, b:

$$I = \ln 2 = 0.693147$$

$$b = \frac{L_s + L_u}{2 \times I \times n \times h} = \frac{359 + 351}{2 \times 0.693147 \times 6 \times 2} = 42.679739$$

Step 3. Calculate the natural logarithm of the estimated potency of the unknown, uncorrected for its assigned potency, M':

$$M' = \frac{y_u - y_s}{b} = \frac{-2.944444}{42.679739} = -0.068989$$

Thus, relative potency of the unknown test solution to that of the standard test solution is given by antiln of –0.068989 as 0.933337 or 93.3%. The estimated potency of the unknown itself can be calculated from its assigned potency, 1500 U/ml, either via the test solution estimated potency as

$$1500 \times 0.933337 = 1400 \text{ U/ml}$$

or directly through M as

$$M = M' + \ln A_U = -0.068989 + \ln 1500 = 7.244231$$

then antiln of 7.244231 = 1400 U/ml

Calculation of Confidence Limits

Step 1. Calculate C as

$$C = \frac{E}{E - s^2 t^2} = \frac{21,004.167}{21,004.167 \times 1.12112 \times 2.06^2} = 1.000227$$

Step 2. Calculate H as

$$H = \frac{E}{b^2 dn} = \frac{21,004.167}{42.6797^2 \times 3 \times 6} = 0.640605$$

and

$$(C-1)(CM'^2 + 2H) = (1.0002 - 1)(1.0002 \times -0.068\,989^2 + 2 \times 0.640605)$$

$$= 0.000257$$

(Note that H is not defined in the pharmacopoeia's Glossary of Symbols at the end of the chapter.)

Step 3. Natural log confidence limits are obtained as

$$\ln A_U + CM' \pm (0.000257)^{0.5} = \ln 1500 + 1.000\,27 \times -0.068989 \pm 0.01603$$

corresponding to confidence limits on the potency estimate of the unknown of

$$= 1378 \text{ to } 1422 \text{ U/ml}$$

Example 9.1 Continued: The Calculation According to United States Pharmacopoeia (USP) Guidance

Equation (16) on p. 1843 of the *USP 24* condenses into one line the information required to obtain s^2, the residual error. That is,

$$s^2 = \left\{ \Sigma y^2 - \Sigma T_r^2 / k - \Sigma T_t^2 / f + T^2 / N \right\} / n \qquad (9.15)$$

where

T = the grand total of observed responses   =   7363
k = the number of column totals   =   6
f = the number of row totals   =   6
n = the number of degrees of freedom for residual error =   25
N = total number of observations   =   36

Now calculate the total sum of squares

$$\Sigma y^2 \qquad\qquad\qquad = 1,527,127$$

then, rows calculation

$$\Sigma T_r^2 / k = \frac{1224^2 + 1239^2 + 1238^2 + 1221^2 + 1226^2 + 1215^2}{6} = 1,506,013.833$$

then, columns (treatments) calculation

$$\Sigma T_t^2 / f = \frac{1057^2 + 1235^2 + 1416^2 + 1043^2 + 1218^2 + 1394^2}{6} = 1,527,023.167$$

$$T^2 / N = 7363^2 / 36 \qquad\qquad = 1,5005,938.028$$

Finally

$$s^2 = (1,527,127 - 1,506,013.833 - 1,527,023.167 + 1,505,938.028)/25 = 1.12112$$

Test of Assay Validity

The first requirement is concerning the value of C, which may be calculated in the same way as described for the *European Pharmacopoeia*. Other requirements concern linearity of the log dose-response lines and their being parallel to one another within the experimental error.

An analysis of variance is carried out using factorial coefficients (also known as orthogonal polynomial coefficients) to calculate sums of deviations squared that are attributable to defined sources. This is illustrated in Table 9.3, which is constructed thus:

TABLE 9.3
Factorial Coefficients for a Three-Dose Level Symmetrical Assay Applied to the Data of Example 9.1 as a Part of the Analysis of Variance

Row	S_1	S_2	S_3	U_1	U_2	U_3	e_i	T_i	$T_i^2/(6 \times e_i)$
a	−1	−1	−1	1	1	1	6	−53	78.03
b	−1	0	1	−1	0	1	4	710	21,004.17
ab	1	0	−1	−1	0	1	4	−8	2.67
q	1	−2	1	1	−2	1	12	4	0.22
aq	−1	2	−1	1	−2	1	12	−2	0.06
Treatment Totals →	1057	1235	1416	1043	1218	1394			

In columns 2 to 7, rows 2 to 6 are the coefficients, which are fixed for this three-dose level symmetrical-assay design.

In row 7, the values for treatment totals are entered.

In column 8, under the heading e_i, are the sums of the squares of the coefficients in each row.

In column 9, under the heading T_i, are the sums of the products of the individual coefficients in each row with their corresponding treatment total.

In column 10, each value is calculated as $(T_i)^2/(6 \times e_i)$

The values in column 10 are the sums of deviations squared corresponding to

$\qquad a$ = difference between standard and unknown

$\qquad b$ = difference between all highest doses and all lowest doses (slopes)

$\qquad ab$ = difference between slopes

$\qquad q$ = common curvature of both preparations

$\qquad aq$ = contrast between curvature of the two preparations

Although the calculation of the sums of deviations squared is identical with that of the *European Pharmacopoeia*, the *United States Pharmacopoeia* interpretation differs. The validity tests utilize the values of $(T_i)^2/(6 \times e_i)$ for rows ab, q, and aq. If the ratio of any of the three values of $(T_i)^2/(6 \times e_i)$ to the residual error s^2 exceeds 3, then compute F_3 thus:

$$F_3 = \Sigma[(T_i)^2/(e_i \times f)]/3s^2 \qquad (9.16)$$

where f is the number of responses to each treatment.

The value of F_3 is then compared with a limiting value from Table 9 of *USP 24* (2000, 1, 844), Section <111> according to the number of degrees of freedom for s^2. The relevant part of that table is reproduced here in Appendix 5.

The function F_3 is applicable to both three-dose level and four-dose level symmetrical assays. The *USP* also gives functions F_1 and F_2, which are applicable, respectively, in the cases of a two-dose level symmetrical assay and an asymmetric assay having three dose levels of one preparation and two dose levels of the other.

In the case of this assay, the ratios are: for ab, q, and aq to s^2 are, respectively,

$$\text{for } ab \quad 2.667/1.121 = 2.38$$

$$\text{for } q \quad 0.222/1.121 = 0.20$$

$$\text{for } ab \quad 0.056/1.121 = 0.05$$

and so there is no requirement to apply any further test. However, solely to illustrate the test, the calculation follows:

$$F_3 = (2.667 + 0.222 + 0.056)/(3 \times 1.121) = 0.876$$

From Table 9 of the *USP 24* Section <111>, the upper limiting value of F_3 for 25 degrees of freedom is 2.99.

Calculation of Potency Estimate

Calculate M', the logarithm of the potency estimate of the unknown uncorrected for its assigned potency as

$$M' = ciT_d/T_b \tag{9.17}$$

where c is a constant obtained from Table 6 in Section <111> of the *USP 24* (2000, 1, 341). For the three-dose level symmetrical-assay design, it has the value 4/3 (compared with Equations (2.5) and (2.6) of Chapter 2, which include, respectively, the fractions 1/4 and 1/3).

i is the logarithm of the ratio between adjacent dose levels, in this case $\log_{10} 2$, T_a and T_b are the values from column 9, rows a and b of Table 9.3.

so that

$$M' = [(4/3) \times 0.30103 \times -53]/710 = -0.02996167$$

Antilog M' is 0.933337, which is identical with the value obtained by the *European Pharmacopoeia* calculation.

Calculation of Confidence Limits

Step 1. Calculate C. This has the same meaning as in the *European Pharmacopoeia*; the calculation procedure is different and so it is shown here. Both procedures lead to exactly the same figure.

$$C = T_b^2 \Big/ \left(T_b^2 - e_b f s^2 f^2 \right) \tag{9.18}$$

where

$T_b = 710$ and is obtained from row b of Table 9.3,

$e_b = 4$ and is obtained from row b of Table 9.3,

$f = 6$ and is the number of responses in each T_i,

s^2 and t have the same meanings and values as before in this example

so that

$$C = 710^2/[710^2 - (4 \times 6 \times 1.12112 \times 2.06^2)] = 1.000226558$$

Step 2. Calculate L, the log confidence interval as

$$L = 2[(C - 1)(CM'^2 + c'i^2)]^{0.5} \tag{9.19}$$

where

$c' = 8/3$ and is a constant for this three-dose level symmetrical assay obtained from Table 6 of the *USP* (2000, 1, 841)

C, M', and i have the same meanings and values as before in this example

so that

$$L = 2\left\{ (1.000227 - 1)\left[(1.000227^2 \times -0.02996167) + (8/3) \times (\log 2)^2 \right] \right\}^{0.5}$$

$$= 2[0.000227 \times 0.242548727]^{0.5} = 0.014\,840\,291$$

Step 3. Calculate the confidence limits as

$$M \pm 0.5L \tag{9.20}$$

where

$$M = M' + \log R \tag{9.21}$$

and R is the assigned potency of 1500 U/ml. Thus

$$M = -0.029962 + 3.176091 = 3.146129$$

and log confidence limits are

$$3.146129 \pm 0.007420 = 3.138709 \text{ to } 3.153549$$

corresponding to confidence limits of 1376.3 to 1424 U/ml ($P = 0.95$).

These figures are expressed to one decimal place to illustrate their coincidence with the figure calculated earlier in this chapter by the *European Pharmacopoeia* procedure. Normally, they would be reported as 1376 to 1424 ($P = 0.95$).

EXAMPLE 9.1 CONTINUED: BACK TO BASICS

The various formulas given in pharmacopoeias ensure that the analyst is given exact instructions for every step in the calculation, thus minimizing the chance of mistakes. Use of such formulas has the disadvantage that the analyst may lose sight of what he or she is really calculating. In the analysis of variance (of Example 9.1 according to the EP guidance, presented earlier in this chapter), each step is the calculation of the sum of the squares of the deviations of the individual total from the mean of those totals, divided by the number of observations that comprise each total. For example:

1. Sum of squares for preparations could have been calculated thus:

		Deviation from Mean	Square of Deviations from Mean
Standard total	3708	26.5	702.25
Unknown total	3655	−26.5	702.25
Mean	3681.5		

Each preparation total is the sum of 18 observations, so the divisor in the next expression is 18.

$$(702.25 + 702.25)/18 = 78.03$$

2. Sum of squares for regression could have been calculated thus:

		Deviation from Mean	Square of Deviations from Mean
High total	2810	355	126,025
Low total	2100	−355	126,025
Mean	2455		

The high totals and low totals are each the sum of 12 observations, so the divisor in the next expression is 12.

$$(126,025 + 126,025)/12 = 21,004.17$$

3. Sum of squares for rows could be calculated in an analagous manner; the divisor in the final expression would be six, the number of observations in a row.

4. Sum of squares for total deviations could be calculated in an analagous manner; the divisor in the final expression would be one because this would be based on 36 individual observations

A THREE-DOSE LEVEL ASSAY FOR ONE UNKNOWN USING A LARGE PLATE AND LATIN SQUARE DESIGN

Recognizing the disadvantages of petri dish assays, Brownlee et al. (1948, 1949) introduced large plates constructed of plate glass. These were leveled using adjustable-height screw feet and a spirit level, thus ensuring a layer of nutrient agar of uniform thickness.

Lees and Tootill (1955a, b, c) introduced balanced statistical designs using a Latin square pattern for the distribution of test solutions on the plate. Such assays are now widely used in Europe.

The assay described now employs a 6 × 6 Latin square design; it provides, in one plate, an assay of one unknown of potentially reasonable precision (each treatment has a replication of 6 corresponding to 18 observations per preparation). It is able to demonstrate adequacy of rectilinearity of the log dose-response lines, thus satisfying *European Pharmacopoeia* criteria for cases when there is not abundant evidence from previous experience of the assay method. The ratio between adjacent dose levels is 2:1, giving an overall dose range for each preparation of 4:1. This is the normal dose ratio used when there are three dose levels; however, an example given in Supplements 2000 and 2001 to the third edition of the *European Pharmacopoeia* uses a 3:2 dose ratio.

Any practicable dose ratio higher than 2:1 would lead to an overall dose range high enough to exacerbate unreasonably the bias due to curvature of the response lines. A bacitracin assay is used to illustrate this design.

EXAMPLE 9.2   AN ASSAY OF BACITRACIN

Test organism: *Micrococcus luteus* NCTC 7743

Design: 6 × 6 Latin square number 12

Standard: house standard BAC002, potency 71.8 IU/mg

Dose ratio: 2:1; nominal dose levels: 1, 2, and 4 IU/ml

Weighing and dilution to high dose standard test solution

$$56.2 \text{ mg} \rightarrow 100 \text{ ml} : 10 \text{ ml} \rightarrow 100 \text{ ml}$$

Calculated potency of high dose standard test solution is 4.03516 IU/ml,

therefore, factor is 4.03516/4 = 1.0088.

Weighing and dilution to high dose unknown test solution

$$57.3 \text{ mg} \rightarrow 100 \text{ ml} : 10 \text{ ml} \rightarrow 100 \text{ ml}$$

Concentration of unknown (a sample of zinc bacitracin) in high-dose test solution is 57.3 µg/ml.

The Latin square design 6 × 6 number LS12 is shown in Figure 9.1. The raw data (zone sizes as appearing on the plate) are shown in Table 9.4 and the unscrambled data are shown in Table 9.5.

4	6	1	5	3	2
3	5	6	4	2	1
2	3	5	1	6	4
6	1	3	2	4	5
5	4	2	3	1	6
1	2	4	6	5	3

FIGURE 9.1 6 × 6 Latin square design number 12.

TABLE 9.4
The Raw Data of Example 9.2, a Three-Dose Level Assay of Bacitracin Using a Large Plate and a 6 × 6 Latin Square Design

	Zone Diameters (mm × 10) as Appearing on the Plate						Row Totals
	236	173	232	205	180	208	1234
	181	207	176	233	209	238	1244
	209	178	207	233	181	237	1245
	181	229	177	207	233	207	1234
	204	229	206	177	231	182	1229
	226	198	220	173	199	174	1190
Column totals →	1237	1214	1218	1228	1233	1246	
Grand total →							7376

TABLE 9.5
Unscrambled Data from Example 9.2, an Assay of Bacitracin and Its Preliminary Inspection

	Zone Diameters (mm × 10)					
	S_1	S_2	S_3	U_1	U_2	U_3
	180	208	232	173	205	236
	181	209	238	176	207	233
	178	209	233	181	207	237
	177	207	229	181	207	233
	177	206	231	182	204	229
	174	198	226	173	199	220
Treatment totals →	1067	1237	1389	1066	1229	1388
Curvature inspection →		170	152		163	159
Slope inspection →		322			322	
Variance →	6.17	17.37	16.30	17.47	9.77	38.67

	Unscrambled Data in Ascending Order of Magnitude					
	S_1	S_2	S_3	U_1	U_2	U_3
	174	198	226	173	199	220
	177	206	229	173	204	229
	177	207	231	176	205	233
	178	208	232	181	207	233
	180	209	233	181	207	236
	181	209	238	182	207	237
Range →	7	11	12	9	8	17
USP G_{1a} →	0.429	0.727	0.250	0.000	0.625	0.529
USP G_{1b} →	0.143	0.000	0.417	0.111	0.000	0.059

The limiting value for G1 is 0.644. This is exceeded in column 2 by the value "0.727". On inspection, it seems that the value "198" of column 2 could be an outlier.

USP R^* → 0.266.

The limiting value for R^* is 0.300 so it appears from this criterion that there is no outlier.

Preliminary Inspection of Data

Referring to Table 9.5:

1. Slopes expressed as the difference in treatment totals, high dose – low dose for both standard and unknown, are identical at 322. Thus the lines are parallel.
2. Considering slopes based on treatment totals first as difference between mid dose and low dose, then as difference between high dose and mid dose, it is seen that the latter is somewhat smaller than the former in the cases of both preparations. This indicates curvature in the direction expected according to the theory of zone formation. It is also in accordance with general experience and is probably no cause for concern.
3. Inspection of the values for variance within each treatment group does not suggest any correlation between the magnitude of variance and dose level. Thus, the assay accords with assumption 4 set forth at the beginning of Chapter 8. The rather large variance of the responses to treatment U_3 is in accordance with this treatment group having the widest range of values, 17. Could the value 220 be an outlier?
4. The two *USP* criteria for detection of outliers are applied. Using a computer program to calculate criterion 1, it is convenient to assume first that the lowest value in a range is the outlier and second that the highest figure is the outlier. Hence the two series of figures in the second part of Table 9.5, G_{1a} and G_{1b}. Referring to Appendix 3, Part A, it is seen that the limiting value for G_1 is 0.644 when $N = 6$. The value for G_1 corresponding to treatment S_2 is 0.727, thus suggesting that the value 198 is an outlier. Now applying criterion 2, R^* is 0.266. Referring to Appendix 3, Part B, it is seen that the limiting value for R^* for six ranges with six values in each range is 0.300. Thus, the finding from criterion 2 does not confirm the suspicions arising from criterion 1.

   Considering all the evidence, the judgment was made that there was not an outlier, and so the calculation was continued without introducing replacement values.

Calculation of Potency Estimate from the Treatment Totals

The expressions for calculation E and F are obtained from the tabulation of Appendix 1. The expressions are related to assay design and replication.

Step 1. Calculate E, the mean increase in response for a twofold increase in dose.

$$E = [(1389 + 1388) - (1067 + 1066)]/(2 \times 2 \times 6) = 26.833$$

There are two contrasts over two dose intervals, each having a replication of six, hence the denominator.

Step 2. Calculate F, the mean difference between unknown and standard responses.

$$F = [(1066 + 1229 + 1388) - (1067 + 1237 + 1389)]/18 = 0.556$$

Step 3. Calculate b, the mean difference in response for a tenfold increase in dose.

$$b = E/\text{I} = 26.833/\log_{10}^2 = 89.1384$$

Step 4. Calculate M, the logarithm of the ratio of the potency of unknown to that of the standard.

$$M = F/b = 0.556/89.1384 = 0.006233$$

Step 5. Calculate potency estimate of high-dose test solution of unknown relative to that of the standard high-dose test solution, then of the unknown itself.

Relative potency estimate of high-dose test solution of unknown is

$$\text{antilog}_{10} \, M = 1.01446$$

therefore, actual potency estimate of test solution is

$$1.01446 \times 4.03516 = 4.0935 \text{ IU/ml}$$

as concentration of test solution is 57.3 µg/ml, the potency of the unknown sample must be

$$4.0935/57.3 = 0.07144 \text{ IU/µg or } 71.44 \text{ IU/mg}$$

Statistical Evaluation

The analysis of variance comprises two parts:

1. Identification of variance due to the six different treatments
2. Identification of variance due to different blocks — rows and columns on the plate in this particular assay design

Variance due to the different individual treatments is calculated using orthogonal polynomial coefficients as shown in Table 9.6.

TABLE 9.6
Factorial Coefficients for a Three-Dose Level Symmetrical Assay Applied to the Data of Example 9.2 as a Part of the Analysis of Variance

Row	S_1	S_2	S_3	U_1	U_2	U_3	e_i	T_i	$T_i^2/(6 \times e_i)$
a	-1	-1	-1	1	1	1	6	-10	2.78
b	-1	0	1	-1	0	1	4	644	17,280.67
ab	1	0	-1	-1	0	1	4	0	0.00
q	1	-2	1	1	-2	1	12	-22	6.72
aq	-1	2	-1	1	-2	1	12	14	2.72

Treatment Totals →	1067	1237	1389	1066	1229	1388

Values for the remaining sources of variation are calculated thus:

Step 1. From the raw data as appearing on the plate calculate the total deviation squares as

$$\left[236^2 + 173^2 + 232^2 + \ldots + 199^2 + 174^2\right] - 7376^2/36 =$$

$$1,529,082 - 7376^2/36 = 17,821.556$$

$$\left[\text{mean correction, } MC = 7376^2/36 = 1,511,260.444\right]$$

Step 2. From the row totals as appearing on the plate calculate

$$\left[1234^2 + 1244^2 + 1245^2 + 1234^2 + 1229^2 + 1190^2\right]/6 - MC$$

$$= 1,511,602.333 - 1,511,260.444 = 341.889$$

Step 3. From the column totals as appearing on the plate calculate

$$\left[\left(1237^2 + 1214^2 + 1218^2 + 1228^2 + 1233^2 + 1246\right]/6 - MC\right.$$

$$= 1,511,379.667 - 1,511,260.444 = 119.223$$

Step 4. From the treatment totals calculate

$$\left[\left(1067^2 + 1237^2 + 1389^2 + 1066^2 + 1229^2 + 1388^2\right)/6\right] - MC$$

$$= 1,528,553.333 - 1,511,260.444 = 17,292.889$$

A summary of the analysis of variance is presented in Table 9.7.

Discussion of Assay Validity Based on the Analysis of Variance

In this assay, the pharmacopoeial requirements concerning variance ratios apply to regression, parallelism, and nonlinearity. The latter has been broken down into its component parts, quadratic curvature and opposed quadratic curvature. Variance ratio for regression at over 5000 is, typically, highly satisfactory. That for parallelism at zero is clearly less than the limit of 4.35. Variance ratios for quadratic and opposed quadratic curvature are both well below the limiting value of 4.35.

Variance ratio for preparations at 0.82 is below the warning level of 4.35 and is consistent with the estimated potency of the unknown, which differs from that of the reference standard by only 1.4%. The values for rows and columns, 20.25 and 7.06, respectively, are way above the warning limit of 2.71. This points to some drift in circumstances across the plate, such as varying thickness of agar medium or uneven temperature of diffusion and/or incubation. It alerts the analyst to review practical operations with a view to reducing such drift. However, the Latin square design automatically tends to correct for row and column differences. The assessment of the analysis of variance is satisfactory and so confidence limits for the potency estimate may be calculated.

TABLE 9.7
Summary of Analysis of Variance, Example 9.2, a Three-Dose Level Assay Using a Large Plate and a 6 × 6 Latin Square Design

Source of Variance	d.f.	Sums of Squares	Mean Squares	Variance Ratio	Limiting Value	At Arbitrary Probability Level, P	Calculated Probability
Preparations	1	2.78	2.78	0.823	<4.35	0.050	0.376
Regression	1	17,280.67	17,280.67	5,116.409	>14.82	0.001	<0.001
Parallelism	1	0.00	0.00	0.000	<4.35	0.050	1.000
Quadratic curvature	1	6.72	6.72	1.990	<4.35	0.050	0.174
Opposed curvature	1	2.72	2.72	0.805	<4.35	0.050˙	0.379
Subtotal	5	17,292.89					
Treatments	5	17,292.89					
Rows	5	341.89	68.38	20.245	<2.71	0.050	<0.001
Columns	5	119.23	23.85	7.060	<2.71	0.050	0.001
Error by difference	20	67.55	3.38	1.000			
Total	35	17,821.56					

Note: Limiting values in column 6 refer to the corresponding arbitrary probability level of column 7.

Step 5. Calculate the term S_{xx}, the sum of the squares of deviations of individual log doses from the mean log dose. The dose ratio is 2:1, so the log dose interval is 0.301. There are 12 high and 12 low doses; mid doses contribute nothing to this calculation, so the value is obtained as

$$S_{xx} = 12 \times [(0.301)^2 + (-0.301)^2] = 3.1744$$

Step 6. Calculate g, the index of significance of the slope, b as

$$g = \frac{s^2 \times t^2}{b^2 \times S_{xx}} \tag{9.22}$$

so that in this assay

$$g = \frac{3.3775 \times 2.086^2}{89.1384^2 \times 3.1774} = 0.000582$$

where s^2, b^2, and S_{xx} have the values already calculated and t is Student's t obtained from tables for $P = 0.95$ and 20 degrees of freedom as 2.086. A brief form of this table is given in Appendix 6.

As g is less than 0.1, calculate confidence limits of the potency estimates by the so-called approximate method.

Step 7. Calculate $V(M')$, the variance of the logarithm of the potency estimate of the unknown, by substituting the known values in the equation

$$V(M) = \frac{s^2}{b^2}\left[\frac{1}{N_s} + \frac{1}{N_T} + \frac{M^2}{S_{xx}}\right] \tag{9.23}$$

$$V(M) = \frac{3.3375}{89.1384^2}\left[\frac{1}{18} + \frac{1}{18} + \frac{0.006233204^2}{3.1774}\right]$$

$$= 0.000425075[0.055555 + 0.055555 + 0.000012] = 0.000047235$$

Step 8. Obtain s_m, the standard error of the logarithm of the potency estimate, as

$$s_m = [V(M)]^{0.5} \tag{9.24}$$

where values are

$$s_m = (0.000047235)^{0.5} = 0.006873$$

Step 9. Log percent confidence limits are given by

$$2 \pm t \times s_m \qquad (9.25)$$

where values are

$$2 \pm 2.086 \times 0.006873 = 1.9857 \text{ and } 2.0143$$

The corresponding percentage limits are 96.8 and 103.4% ($P = 0.95$).
Alternatively, expressed in terms of the standard potency, the estimated potency is 101.4% with limits 98.2 to 104.9% ($P = 0.95$).

A TWO-DOSE LEVEL ASSAY FOR FOUR PREPARATIONS (THREE UNKNOWNS) USING A LARGE PLATE AND AN 8 × 8 LATIN SQUARE DESIGN

This assay, using an 8 × 8 Latin square design, provides for the simultaneous assay of three unknowns at two dose levels. Each treatment has a replication of eight, corresponding to 16 observations per preparation. This design may be useful in routine quality control in the pharmaceutical industry when levels of "assigned potency" (before assay) may be given with reasonable confidence. The dose ratio employed is frequently 2:1 or 4:1. In this case, the ratio chosen was 5:1, no doubt because the generally poor slope of neomycin assays permits a wider range to be used.

EXAMPLE 9.3 AN ASSAY OF NEOMYCIN

Test organism: *Bacillus pumilis* NCTC 8241

Design: 8 × 8 Latin square 17

Standard: house standard neo04 potency 730 IU/mg

Dose ratio: 5:1; nominal dose levels: 10 and 2 IU/ml

Weighing and dilution to high-dose standard test solution

$$34.9 \text{ mg} \rightarrow 250 \text{ ml} : 10 \text{ ml} \rightarrow 100 \text{ ml}$$

Calculated potency of high dose standard test solution is 10.19 IU/ml, therefore, factor is 1.019.

Three unknown preparations included:

> Two batches of water-soluble creams having a stated content of 3500 IU/g of neomycin activity
> One batch of tablets having a stated content of 350,000 IU/tablet of neomycin activity

The weighing and dilution procedures were:

Sample 1

Mix the cream well and weigh approximately 1.4 g and dilute thus:

$$1.4326 \text{ g} \rightarrow 500 \text{ ml} : 10 \text{ ml} \rightarrow 100 \text{ ml} : 10 \text{ ml} \rightarrow 100 \text{ ml}$$

This gave a concentration of cream in the test solution of 0.000028652 g/ml.

Sample 2

Mix the cream well and weigh approximately 1.4 g and dilute thus:

$$1.3977 \text{ g} \rightarrow 500 \text{ ml} : 10 \text{ ml} \rightarrow 100 \text{ ml} : 10 \text{ ml} \rightarrow 100 \text{ ml}$$

This gave a concentration of cream in the test solution of 0.000027954 g/ml.

Sample 3

Determine the average weight of a tablet; crush 20 tablets and mix well; take a weight of powdered tablets that should contain about 100,000 units of neomycin activity.

The average weight of a tablet was 593.2 mg, so that 100,000 units of neomycin activity should be contained in about 169 mg of powder.

The actual weighing and dilutions were:

$$160.3 \text{ mg} \rightarrow 500 \text{ ml} : 5 \text{ ml} \rightarrow 100 \text{ ml}$$

This gave a concentration of powdered tablets in the test solution of 0.01603 mg/ml

The Latin square design 8 × 8 number LS17 is shown in Figure 9.2.

The raw data (zone diameters × 10) as appearing on the assay plate are shown in Table 9.8, together with column totals and row totals (which are used in the analysis of variance).

The unscrambled raw data, treatment observations arranged in columns, are shown in Table 9.9 together with various figures calculated from these data.

Preliminary Inspection of Data

Referring first to Table 9.8, the variation between row totals appears unremarkable; similarly the variation between column totals appears unremarkable. There seems to be no concern about assay quality on these counts.

Now turning to Table 9.9:

4	1	8	7	6	5	2	3
1	4	5	6	7	8	3	2
8	5	4	3	2	1	6	7
3	2	7	8	5	6	1	4
2	3	6	5	8	7	4	1
5	8	1	2	3	4	7	6
6	7	2	1	4	3	8	5
7	6	3	4	1	2	5	8

FIGURE 9.2 8 × 8 Latin square design number 17 (as used in the neomycin assay, example 9.4.).

TABLE 9.8
The Raw Data from Example 9.3, a Two-Dose Level Assay for Three Unknowns Using an 8 × 8 Latin Square Design

Zone Diameters (mm × 10) as Appearing on the Assay Plate								Row Totals
213	253	214	253	216	249	217	250	1865
244	213	238	207	246	209	246	213	1816
210	242	206	245	209	246	208	250	1816
243	211	244	208	243	210	248	213	1820
214	244	212	245	214	250	213	251	1843
246	209	246	212	251	213	247	214	1838
210	250	214	251	212	250	214	251	1852
249	210	246	212	250	212	252	217	1848
Column Totals →								
1829	1832	1820	1833	1841	1839	1845	1859	

1. Ranges of observations within treatments are very close with the exception of treatment U_{3h}, but based on general experience, this is not considered cause for concern. Similarly, for that treatment, variance is rather high but not considered cause for concern.
2. Individual preparation slopes expressed as differences between their high- and low-dose totals all lie within a narrow range, so there is no cause to suspect deviations from parallelism.
3. The tests for outliers provide conflicting guidance. Applying *USP* criterion 1, the value for G_2 is 0.857 for treatment U_{2l}, whereas the limiting value in these circumstances (obtained from Table 8.1) is 0.780. This suggests that the value 206 is an outlier. However, applying *USP* criterion 2 to the ranges as a whole the value of R^* is calculated as 0.192, which is less than the limiting value 0.218 (obtained from Appendix 3).

TABLE 9.9
The Unscrambled Data of Example 9.3, a Two-Dose Level Neomycin Assay of Three Unknowns and Its Preliminary Inspection

			Zone Diameters (mm × 10)					Row
U_{1H}	U_{1L}	U_{2H}	U_{2L}	U_{3H}	U_{3L}	S_H	S_L	Totals
253	217	250	213	249	216	253	214	1865
244	213	246	213	238	207	246	209	1816
246	209	245	206	242	208	250	210	1816
248	211	243	213	243	210	244	208	1820
251	214	244	213	245	212	250	214	1843
246	212	251	213	246	214	247	209	1838
251	214	250	212	251	210	250	214	1852
250	212	246	212	252	210	249	217	1848

Treatment
totals (H) → 1989		1975		1966		1989		7919
totals (L) →	1702		1695		1687		1695	6779

Preparation
totals →   3691   3670   3653   3684

Ranges →	9	8	8	7	14	9	9	9
USP G_{2a}	0.286	0.600	0.286	0.857	0.385	0.200	0.500	0.167
USP G_{2b}	0.286	0.500	0.143	0.000	0.300	0.750	0.429	0.375

The limiting value for G_2 is obtained from Appendix 3, part B as 0.780. As this value is exceeded in column U_{2L}, this suggests that the value 206 is an outlier.

USP R^* →   0.192

The limiting value for criterion 2 is obtained from Table 8.2 as 0.218 and so, on this basis, there is no reason to suspect an outlier.

Slope
inspection →   287   280   279   294

Variance →   9.696   5.643   9.268   5.839   22.786   8.982   7.982   10.696

    In consideration of the narrow range of observations in treatment column U_{2l}, the judgment was reached that it does not include an outlier. The value 206 is quite consistent with the other values in this treatment column. The reason for the alarm triggered by *USP* criterion 1 is clearly that all values except 206 are clustered together at the high end of the range, thus leaving a large gap between 206 and its nearest neighbor.

    In conclusion, although the value of these statistical criteria must be recognized, so must their fallibility. This is allowed for in the probability statements.

On the basis of this preliminary inspection there is no reason to doubt the soundness of the raw data, so it is appropriate to proceed with the routine calculation of potency estimate and statistical evaluation.

Calculation of Potency Estimate from the Treatment Totals

The expressions for calculation of E and F are obtained from the tabulation of Appendix 1. The expressions are related to assay design and replication.

Step 1. Calculate of E, the mean increase in response for a fivefold increase in dose.

$$E = (7919 - 6779)/(2 \times 16) = 35.625$$

Step 2. Calculate the three values of F, the mean differences in response between the three unknowns and the standard.

$$F_1 = (3691 - 3684)/(2 \times 8) = +0.4375$$

$$F_2 = (3670 - 3684)/(2 \times 8) = -0.8750$$

$$F_3 = (3653 - 3684)/(2 \times 8) = -1.9375$$

Step 3. Calculate b, the mean difference in response for a tenfold increase in dose.

$$b = E/I = 35.625/\log_{10} 5 = 50.9679$$

Step 4. Calculate the three values of M, the logarithms of the ratios of the potency of unknowns to that of the standard.

$$M_1 = F_1/b = 0.4375/50.9679 = 0.008584$$

$$M_2 = F_2/b = -0.8750/50.9679 = -0.017168$$

$$M_3 = F_3/b = -2.3125/50.9679 = -0.038014$$

Step 5. Calculate potency estimate of high-dose test solutions of unknowns relative to that of the standard high-dose test solution.

Relative potency estimate of high-dose test solutions of unknowns are

$$\text{antilog}_{10} M_1 = 1.0200$$
$$\text{antilog}_{10} M_2 = 0.9612$$
$$\text{antilog}_{10} M_3 = 0.9162$$

Step 6. Calculate actual potency of the samples:

Sample 1

Potency estimate of the high-dose test solution is

$$1.0200 \times 1.019 \times 10 = 10.35 \text{ IU/ml}$$

Concentration of cream in the high-dose test solution is 0.000028652 g/ml, therefore, estimated concentration of neomycin activity in the cream is

$$10.35/0.000028652 \text{ IU/g} = 361231 \text{ IU/g}$$

which is 103.2 percent of the claimed potency.

Sample 2

Potency estimate of the high-dose test solution is

$$0.9612 \times 1.019 \times 10 = 9.795 \text{ IU/ml}$$

Concentration of cream in the high-dose test solution is 0.000027954 g/ml, therefore, estimated concentration of neomycin activity in the cream is

$$9.795/0.000027954 \text{ IU/g} = 350397 \text{ IU/g}$$

which is 100.1 percent of the claimed potency.

Sample 3

Potency estimate of the high-dose test solution is

$$0.9162 \times 1.019 \times 10 = 9.336 \text{ IU/ml}$$

Concentration of powdered tablet in the high-dose test solution is 0.01603 mg/ml, therefore, estimated concentration of neomycin activity in the powdered tablet is

$$9.336/0.01603 \text{ IU/mg} = 582.4 \text{ IU/mg}$$

and estimated neomycin activity in a tablet of average weight (593.2 mg) is

$$593.2 \times 582.4 = 345,487 \text{ IU/tablet}$$

which is 98.7 percent of the claimed potency.

Statistical Evaluation

Step 1. From the raw data as appearing on the plate, calculate the total deviation squares as

$$\left[213^2 + 253^2 + 214^2 + \ldots + 152^2 + 217^2\right] - 14,698^2/64 =$$

$$3,396,422 - 3,375,487.56 = 20,394.44$$

$$\left(\text{mean correction, } MC = 14,698^2/64 = 3,375,487.56\right)$$

Step 2. From the row totals as appearing on the plate, calculate

$$\left[1865^2 + 1816^2 + 1816^2 + 1820^2 + 1843^2 + 1838^2 + 1852^2 + 1848^2\right]/8 - MC$$

$$= 3,375,779.75 - 3,375,487.56 = 292.19$$

Step 3. From the column totals as appearing on the plate, calculate

$$\left[\left(1829^2 + 1832^2 + 1820^2 + 1833^2 + 1841^2 + 1839^2 + 1845^2 + 1859^2\right]/8 - MC\right.$$

$$= 3,375,607.25 - 3,375,487.56 = 120.19$$

Step 4. From the treatment totals, calculate

$$\left[\left(1989^2 + 1702^2 + 1975^2 + 1695^2 + 1,66^2 + 1687^2 + 1989^2 + 1695^2\right)/8\right] - MC$$

$$= 3,395,855.75 - 3,375,487.56 = 20,268.19$$

Step 5. From the preparation totals, calculate

$$\left[\left(3691^2 + 3670^2 + 3653^2 + 3684^2\right)/16\right] - MC$$

$$= 3,375,540.38 - 3,375,487.56 = 52.82$$

Step 6. From the high- and low-dose totals, calculate regression squares as

$$(7919 - 6779)^2/64 = 20,306.25$$

Step 7. From the high- and low-dose totals for individual preparations, calculate parallelism squares as

$$\left\{\left[(1989-1702)^2 + (1975-1695)^2 + (1966-1687)^2 + (1989-1695)^2\right]\Big/16\right\} -$$

regression squares

$$= 20,315.38 - 20,306.25 = 9.125$$

Discussion of Assay Validity Based on the Analysis of Variance

The results of the analysis of variance are summarized in Table 9.10. An interpretation of the variance ratios in that table follows.

In this two-dose level assay, the pharmacopoeial requirements concerning variance ratios apply to regression and parallelism only. The variance ratio for regression at over 5000 greatly exceeds the limiting value of 12.6, which was obtained from the 0.1% table for $n_1 = 1$ and $n_2 = 41$ such as in Appendix 4. The variance ratio for parallelism at 0.831 is well below the limiting value of 2.83.

Now concerning nonpharmacopoeial criteria the value for preparations at 4.806 exceeds the warning level of 2.83. This is apparently due to unknown 3, which has an estimated potency ratio of 0.90. At that estimated potency, and in consideration of the relatively wide dose ratio (5:1), there might be some slight bias due to

---

TABLE 9.10
Summary of Analysis of Variance for Example 9.3 a Neomycin Assay for Three Unknowns

Source of Variance	d.f.	Sum of Squares	Mean Squares	Variance Ratio	Limiting Value	At Arbitrary Probability Level, *P*	Calculated Probability
Preparations	3	52.82	17.61	4.806	<2.83	0.050	0.007
Regression	1	20,306.25	20,306.25	5542.747	>12.6	0.001	<0.001
Parallelism (deviations from)	3	9.13	3.04	0.831	<2.83	0.050	0.494
Sub total	7	20,368.20					
Treatments	7	20,368.19					
Rows	7	292.19	41.74	11.394	<2.25	0.050	<0.001
Columns	7	120.19	17.17	4.687	<2.25	0.050	0.001
Error by difference	42	153.87	3.66	1.000			
Total	63	20,934.44					

Limiting values in column 6 refer to the corresponding arbitrary probability level of column 7.

curvature, which may well exist, although it cannot be detected in this assay. However, this is not considered to be serious enough to invalidate the assay. See also the discussion on the influence of curvature in Chapter 11. The variance ratios for rows and columns are quite significant and point to some drift in circumstances across the plate. However, the Latin square design automatically tends to correct for such differences. The assessment of the analysis of variance is satisfactory and so confidence limits for the potency estimate may be calculated.

Calculate the term S_{xx}, the sum of the squares of deviations of individual log doses from the mean log dose. The dose ratio is 5:1, so the log dose interval is 0.69897. There are 32 high and 32 low doses, so the value is obtained as:

$$32[(0.69897/2)^2 + (-0.69897/2)^2] = 7.816945$$

Calculate g, the index of significance of the slope, b, using Equation (9.22) as

$$g = \frac{3.7486 \times 2.020^2}{50.9679^2 \times 7.8169} = 0.000753$$

where s^2, b^2, and S_{xx} have the values already calculated and t is Student's t obtained from tables for $P = 0.95$ and 42 degrees of freedom as 2.020 as in Appendix 6.

Because g is less than 0.1, calculate confidence limits of the potency estimates by the so-called approximate method.

First calculate $V(M_1)$ the variance of the logarithm of the potency estimate of the first unknown, by substituting the known values in Equation (9.23).

$$V(M_1) = \frac{3.7486}{50.2361^2}\left[\frac{1}{16} + \frac{1}{16} + \frac{0.008584^2}{7.8169}\right]$$

$$= 0.001485378[0.0625 + 0.0625 + 0.000007] = 0.000185781$$

The standard error of the logarithm of the potency estimate, s_M, is then obtained by Equation (9.24) as:

$$s_M = (0.000185781)^{0.5} = 0.01363018$$

Log percent confidence limits are given by Equation (9.25) as:

$$2 \pm 2.020 \times 0.01363018 = 1.9725 \text{ and } 2.0274$$

The corresponding percentage limits are 93.9 and 106.5% ($P = 0.95$). Similarly, almost identical percentage confidence limits were calculated for unknowns 2 and 3.

A TWO-DOSE LEVEL ASSAY FOR TWO PREPARATIONS (ONE UNKNOWN) USING A LARGE PLATE AND 8 × 8 LATIN SQUARE DESIGN WITH TWO WEIGHINGS OF EACH PREPARATION

In all assay designs hitherto discussed in this book, there has been only one weighing for each preparation. It is implicit in the evaluation of those assays that test solutions have been prepared accurately and that there is no error in the x axis of the log dose-response line. The statistical evaluation then provides a measure of the random error in the responses — the y axis. Such assays will now be designated (in this book) as class 1 assays.

The assay design under consideration now (designated here as a class 2 assay) differs from those not including more than one weighing per preparation in an important principle. In class 2 assays, the statistical evaluation is intended to provide also a measure of any differences between the test solutions arising from errors in the two weighings from each preparation (or in dilution errors).

It follows that the analyst must endeavor to prepare the two primary solutions from each preparation at exactly the same actual strength. The same nominal strength is not acceptable.

The special feature of the evaluation of this assay is in the analysis of variance, where mean squares are calculated for deviations due to differences in the actual concentrations of test solutions, which are intended to be identical.

The origin of this design is not known to the writer. In its original form the seven degrees of freedom for treatments were broken down thus:

Source of Variation	d.f.
Preparations	1
Regression	1
Parallelism	1
Weighings within preparations	2
Regression × weighings	2

The meanings of the terms *weighings within preparations* and *regression × weighings* are not immediately self-evident to the analyst, nor is their individual relevance self-evident. What concerns the analyst is only that errors in weighings/dilutions that are big enough to make an unacceptable error in the potency estimate can be detected.

In Example 9.4a, the original form of analysis of variance is followed. In a development of the calculation (Example 9.4b), *weighings within preparations* and *regression × weighings* are broken down to their component parts, each having one degree of freedom.

Example 9.4a   An Assay of Neomycin

Test organism: *Bacillus pumilis* NCTC 8241

Design: Latin square design number LS 02

Standard: house standard potency 730 IU/mg

Dose ratio: 5:1; nominal dose levels: 10 and 2 IU/ml

Weighing and dilution to high-dose standard test solutions

Solution A:

$$35.3 \text{ mg} \rightarrow 250 \text{ ml} : 10 \text{ ml} \rightarrow 100 \text{ ml}$$

The potency of high dose standard test solution A is calculated as:

$$(35.3 \times 730 \times 10)/(250 \times 100) = 10.31 \text{ IU/ml}$$

Solution B

Because it is not easy to weigh exactly the same weight of standard for the second solution, it is convenient to weigh slightly more for solution B, then calculate the required volume for the primary dilution. The amount of standard weighed was 35.8 mg, that is 101.416 percent of that for standard A. Thus, the initial dilution had to be to 101.416 percent of 250 ml, that is, 253.5 ml.

Thus, in this case the neomycin standard was first diluted to 250 ml, then 3.5 ml of diluent was measured by pipette and added to the solution. It was, of course, essential to ensure that there was thorough mixing after this addition.

$$35.8 \text{ mg} \rightarrow 253.5 \text{ ml} : 10 \text{ ml} \rightarrow 100 \text{ ml}$$

Weighing and dilution to high-dose unknown test solution A:

$$37.7 \text{ mg} \rightarrow 250 \text{ ml} : 10 \text{ ml} \rightarrow 100 \text{ ml}$$

Concentration of the unknown in high-dose test solution A is calculated as

$$(37.7 \times 1000 \times 10)/(250 \times 100) = 15.08 \text{ µg/ml}$$

Weighing and dilution to high-dose unknown test solution B: 38.2 mg of unknown was weighed; it was calculated that this should be diluted to 253.3 ml

$$38.2 \text{ mg} \rightarrow 253.3 \text{ ml} : 10 \text{ ml} \rightarrow 100 \text{ ml}$$

Concentration of the unknown in high-dose test solution B is also 15.08 µg/ml

The 8 × 8 Latin square design 02 is shown in Figure 9.3.

The raw data (zone diameters mm) as appearing on the assay plate are shown in Table 9.11, together with column totals and row totals (which are used in the analysis of variance) and the grand total. The unscrambled raw data, treatment observations arranged in columns, are shown in Table 9.12, together with various figures calculated from these data.

Preliminary Inspection of Data

Referring first to Table 9.11, the variation between row totals appears to be typical of what may be expected of such an assay; similarly, the variation between column totals appears to be acceptable. There seems to be no concern about assay quality on these counts.

6	3	2	1	8	7	4	5
3	6	7	8	1	2	5	4
2	7	6	5	4	3	8	1
5	4	1	2	7	8	3	6
4	5	8	7	2	1	6	2
7	2	3	4	5	6	1	8
8	1	4	3	6	5	2	7
1	8	5	6	3	4	7	2

FIGURE 9.3 8 × 8 Latin square design number 02.

TABLE 9.11
The Raw Data From Example 9.4, a Two-Dose Level Assay for One Unknown Using an 8 × 8 Latin Square Design With Two Weighings for Each of the Two Preparations

Zone Diameters (mm × 10) as Appearing on the Assay Plate								Row Totals
23.4	26.0	23.2	26.0	23.4	26.3	23.4	26.1	197.8
25.9	23.0	25.8	23.2	26.0	23.0	26.1	23.0	196.0
23.0	26.1	23.4	26.1	23.0	25.8	23.4	26.2	197.0
25.8	22.8	25.9	23.0	26.1	23.4	25.8	23.4	196.2
23.2	25.9	23.1	26.1	23.3	26.1	23.2	25.9	196.8
26.1	22.9	26.1	23.0	26.2	22.6	25.6	23.0	195.5
23.3	26.0	22.8	25.8	23.3	25.9	22.8	26.0	195.9
26.2	23.0	26.0	23.3	25.8	22.8	26.2	23.4	196.7
Column totals → 196.9	195.7	196.3	196.5	197.1	195.9	196.5	197.0	

TABLE 9.12
The Unscrambled Data From Example 9.4, a Two-Dose Level Assay of Neomycin with Two Weighings of Each of Standard and Unknown, Together With Its Preliminary Inspection

Treatment Test Solution Number	Zone Diameters (mm)							
	SA_H	SA_L	SB_H	SB_L	UA_H	UA_L	UB_H	UB_L
	1	2	3	4	5	6	7	8
	26.0	23.2	26.0	23.4	26.1	23.4	26.3	23.4
	26.0	23.0	25.9	23.0	26.1	23.0	25.8	23.2
	26.2	23.0	25.8	23.0	26.1	23.4	26.1	23.4
	25.9	23.0	25.8	22.8	25.8	23.4	26.1	23.4
	26.1	23.3	25.9	23.2	25.9	23.2	26.1	23.1
	25.6	22.9	26.1	23.0	26.2	22.6	26.1	23.0
	26.0	22.8	25.8	22.8	25.9	23.3	26.0	23.3
	26.2	23.4	25.8	22.8	26.0	23.3	26.2	23.0
Treatment totals	208.0	184.6	207.1	184.0	208.1	185.6	208.7	185.8
Treatment means	26.00	23.08	25.89	23.00	26.01	23.20	26.09	23.23
Slope inspection →		2.9		2.9		2.8		2.9
Variance →	0.037	0.042	0.013	0.046	0.018	0.077	0.021	0.031
Range →	0.6	0.6	0.3	0.6	0.4	0.8	0.5	0.4

Now turning attention to Table 9.12, the individual slopes are very close to one another, also the corresponding A and B treatment totals appear to be reasonably close to one another. However, see the comment in the section, "A More Critical Examination of an Assay with Duplicate Weighings," later in this chapter.

Calculate Potency Estimate from the Treatment Totals

The expressions for calculation E and F are obtained from the tabulation of Appendix 1. The expressions are related to assay design and replication

Step 1. Calculate E, the mean increase in response for a fivefold increase in dose.

$$E = (831.9 - 740.0)/(2 \times 16) = 2.871875$$

where 831.9 is the total of responses to all four high-dose treatments and 740.0 is the total of responses to all four low-dose treatments.

Step 2. Calculate the value of F, the mean difference in response between all unknown and all standard treatments.

$$F = (788.2 - 783.7)/(2 \times 16) = + 0.140625$$

Step 3. Calculate b, the mean difference in response for a tenfold increase in dose.

$$b = E/I = 2.871875/\log_{10} 5 = 4.10872424$$

Step 4. Calculate the value of M, the logarithm of the ratio of the potency of unknown to that of the standard.

$$M = F/b = + 1.40625/4.10872424 = 0.034225952$$

Step 5. Calculate potency estimate of high-dose test solutions of unknown relative to that of the standard high-dose test solutions. Relative potency estimates of high-dose test solutions of unknowns are

$$\text{antilog}_{10} M = 1.082$$

Step 6. Calculate actual potency estimates of test solution as

$$10.31 \times 1.082 \text{ IU/ml} = 11.155 \text{ IU/ml}$$

Concentration of neomycin sulphate in unknown test solutions is 15.08 μg/ml or 0.01508 mg/ml, therefore, potency of neomycin sulphate is

$$11.155/0.01508 = 739.7 \text{ IU/mg}$$

Statistical Evaluation

Step 1. From the raw data as appearing on the plate, calculate the total deviation squares as

$$23.4^2 + 26.0^2 + 23.2^2 + \ldots + 22.8^2 + 26.2^2 + 23.4^2 - 1{,}571.9^2/64$$

$$38{,}741.7 - 38{,}607.34 = 134.36$$

(where $1571.9^2/64 = 38{,}607.34$ is the mean correction or MC)

Step 2. From the row totals as appearing on the plate, calculate

$$\left[197.8^2 + 196.0^2 + 197.0^2 + 196.2^2 + 196.8^2 + 195.5^2 + 195.9^2 + 196.7^2\right]/8 - MC$$

$$= 38{,}607.8088 - 38{,}607.3400 = 0.4711$$

Step 3. From the column totals as appearing on the plate, calculate

$$\left[\left(196.9^2 + 195.7^2 + 196.3^2 + 196.5^2 + 197.1^2 + 195.9^2 + 196.5^2 + 197.0^2\right)\big/8 - MC\right.$$

$$= 38,607.5638 - 38,607.3400 = 0.2238$$

Step 4. From the treatment totals, calculate

$$\left[\left(208.1^2 + 184.6^2 + 207.1^2 + 184.0^2 + 208.1^2 + 185.6^2 + 208.7^2 + 185.8^2\right)\big/8\right] - MC$$

$$= 38,739.7338 - 38,607.3400 = 132.3961$$

Step 5. From the preparation totals, calculate

$$\left[\left(208.0 + 184.6\right)^2 + \left(207.1 + 184.0\right)^2 + \left(208.1 + 185.6\right)^2 + \left(208.7 + 185.8\right)^2\big/32\right] - MC$$

$$= 38,607.6541 - 38,607.3400 = 0.3164$$

Step 6. From the high- and low-dose totals, calculate regression squares as

$$[(208.0 + 207.1 + 208.1 + 208.7) - (184.6 + 184.0 + 185.6 + 185.8)]^2/64$$

$$= 131.9627$$

Step 7. From the high- and low-dose totals for individual preparations, calculate parallelism squares as

$$\left\{\left(\left[(208.0 + 207.1) - (184.6 + 184.0)\right]^2 + \left[(208.1 + 208.7) - (185.6 + 185.8)\right]^2\right)\big/32\right\}$$

$$- \text{regression squares} = 131.9816 - 131.9627 = 0.0189$$

Step 8. Calculate weighings within preparations as

$$\left\{\left[\left(SA_H + SA_L\right) - \left(SB_H + SB_L\right)\right]^2\big/32\right\} + \left\{\left[\left(UA_H + UA_L\right) - \left(UB_H + UB_L\right)\right]^2\big/32\right\}$$

$$= \left\{\left[(208.0 + 184.6) - (207.1 + 184.0)\right]^2\big/32\right\}$$

$$+ \left\{\left[(208.1 + 185.6) - (208.9 + 185.8)\right]^2\big/32\right\} = 0.0903$$

Step 9. Calculate regression × weighings as

It is convenient to express this lengthy calculation as regression × weighings = A – B – C.

where

$$A = [(SA_H - SA_L)^2/16] + [(SB_H - SB_L)^2/16] + [(UA_H - UA_L)^2/16] + [(UB_H - UB_L)^2/16]$$

$$B = \text{regression squares for standard} = \{[(SA_H + SB_H) - (SA_L + SB_L)]^2\}/32$$

$$C = \text{regression squares for unknown} = \{[(UA_H + UB_H) - (UA_L + UB_L)]^2\}/32$$

so that

$$A = [(208.0 - 184.6)^2/16] + [(207.1 - 184.0)]^2/16\} + [(208.1 - 185.6)^2/16]$$
$$+ [(208.7 - 185.8)]^2/16\} = 131.9894$$
$$B = \{[(208.1 + 207.1) - (184.6 + 184.0)]^2\}/32 = 67.5703$$
$$C = \{[(208.1 + 208.7) - (185.6 + 185.8)]^2\}/32 = 64.4113$$

and regression × weighings = 131.9894 – 67.5703 – 64.4113 = 0.0078

TABLE 9.13A

Summary of the Analysis of Variance for Example 9.4a, a Two-Dose Level Assay Comparing One Unknown With the Reference Standard Using a Large Plate and an 8 × 8 Latin Square Design, with Duplicate Weighings of Each of the Two Preparations

Source of Variance	d.f.	Sum of Squares	Mean Squares	Variance Ratio	Limiting Value	At Arbitrary Probability Level, *P*	Calculated Probability
Preparations	1	0.316	0.316	10.230	<4.06	0.050	0.003
Regression	1	131.963	131.963	4266.702	>12.6	0.001	<0.001
Parallelism [deviations from]	1	0.019	0.019	0.611	<4.06	0.050	0.433
Weighings within preparations	2	0.090	0.045	1.460	<3.22	0.050	0.244
Regression × weighings	2	0.008	0.004	0.126	<3.22	0.050	0.878
Subtotal	7	132.396					
Treatments	7	132.396					
Rows	7	0.471	0.067	2.176	<2.20	0.050	0.057
Columns	7	0.226	0.032	1.044	<2.20	0.050	0.418
Error by difference	42	1.299	0.031	1.000			
Total	63	134.392					

Note: Limiting values in column 6 refer to the corresponding arbitrary probability level of column 7.

Discussion of Assay Validity Based on the Analysis of Variance

In this two-dose level assay, the pharmacopoeial requirements concerning variance ratios apply to regression and parallelism only. The variance ratio for regression at over 4000 greatly exceeds the limiting value of 12.4. The variance ratio for parallelism at 0.61 is well below the limiting value of 4.06.

The value for preparations at 10.23 exceeds the warning level of 4.06. This is consistent with the estimated potency of the unknown being 108.2 percent of that of the standard. For the same reasons as given in Example 9.2, it is considered that there may be a slight bias in the potency estimate but not enough to invalidate the assay. The values for variance ratios for weighings within preparations and regression × weighings are less than the arbitrary limit so these seem satisfactory. (However, a different conclusion is reached from the more critical review in Example 9.4b).

As the assessment of the analysis of variance seems satisfactory, confidence limits for the potency estimate may be calculated.

Step 10. Calculate the term S_{xx}, the sum of the squares of deviations of individual log doses from the mean log dose. The dose ratio is 5:1, so the log dose interval is 0.69897. There are 32 high and 32 low doses, so the value is obtained as:

$$32[(0.698970/2)^2 + (-0.698970/2)^2] = 7.816945$$

Step 11. Calculate g, the index of significance of the slope, b, using Equation (9.22) as

$$g = \frac{0.03091 \times 2.020^2}{4.1087^2 \times 7.816945} = 0.0009564$$

The value of g is well below the limit of 0.1 and so the variance of M and confidence limits may be calculated by the approximate method.

Step 12. Calculate the variance of M by Equation (9.23) as

$$V(M) = \frac{0.030931}{4.1087^2}\left[\frac{1}{32} + \frac{1}{32} + \frac{0.0342^2}{7.816945}\right]$$

$$= 0.00183225\,(0.03125 + 0.03125 + 0.000149628) = 0.000114789$$

Step 13. Calculate s_M by Equation (9.24) as

$$s_M = (0.000114789)^{0.5} = 0.0107140$$

Step 14. Log percent confidence limits are

$$2 \pm 0.0107140 \times 2.020 = 1.978358 \text{ to } 2.021642$$

corresponding to confidence limits of 95.1 to 105.1%.

EXAMPLE 9.4B A MORE CRITICAL EXAMINATION OF AN ASSAY WITH DUPLICATE WEIGHINGS

The raw data of Example 9.4 are examined here in Example 9.4b in more detail to gain an insight into the meanings of the components of variation attributed to differences in the two weighings of standard and unknown. A simple comparison of standard *B* with standard *A* and similarly a comparison of unknown *B* with unknown *A* will give figures that are immediately meaningful to the analyst. Using the value of *b* obtained from a consideration of all 64 responses (4.1087), calculate the potency of standard *B* in terms of standard *A*:

$$F = \left[(207.1 + 184.0) - (208.0 + 184.6)\right]/16 = -0.09375$$

$$M = F/b = -0.09375/4.1087 = -0.02282$$

$$R = \text{antilog} \; -0.02282 = 0.9488$$

Similarly, calculate the potency of unknown *B* in terms of unknown *A*:

$$F = \left[(208.7 + 185.8) - (208.1 + 185.6)\right]/16 = 0.0500$$

$$M = F/b = 0.0500/4.1087 = 0.01217$$

$$R = \text{antilog} \; 0.01217 = 1.0284$$

In round figures the discrepancies between weighings *B* and weighings *A* are −5% in the case of the standards and +3% in the case of the unknowns. This is clearly not good enough in an assay intended to be of quite high accuracy and precision. Yet the statistical test for significance of differences due to weighing has not alerted us to this unsatisfactory situation.

The two components of variation (1) weighings within preparations, and (2) regression × weighings are replaced here by four components: (1) becomes (1a) contrast of standards + (1b) contrast of unknowns; (2) becomes (2a) contrast of standard slopes + (2b) contrast of unknown slopes.

These four, which are much more meaningful expressions, each have one degree of freedom.

Calculate:

1a. Contrast of standards as

$$\left[\left(SA_H + SB_L\right) - \left(SB_H + SB_L\right)\right]^2 \big/ 32$$

$$= \left[\left(208.0 + 184.6\right) - \left(207.1 + 184.0\right)\right]^2 \big/ 32 = 0.0703$$

1b. Contrast of unknowns as

$$\left[\left(UA_H + UB_L\right) - \left(UB_H + UB_L\right)\right]^2 \big/ 32$$

$$= \left[\left(208.1 + 185.6\right) - \left(208.7 + 185.8\right)\right]^2 \big/ 32 = 0.0200$$

2a. Contrast of standard slopes as

$$\left\{\left[\left(SA_H - SA_L\right)^2\right] \big/ 16\right\} + \left\{\left[\left(SB_H - SB_L\right)^2\right] \big/ 16\right\}$$

– regression squares for standard

where regression squares for standard is

$$\left[\left(SA_H + SB_H\right) - \left(SA_L + SB_L\right)\right]^2 \big/ 32$$

Thus contrast of standard slopes is

$$\left\{\left[\left(208.0 - 184.6\right)^2\right] \big/ 16\right\} + \left\{\left[\left(207.1 - 184.0\right)^2\right] \big/ 16\right\}$$

$$-\left[\left(208.0 + 207.1\right) - \left(184.6 + 184.0\right)\right]^2 \big/ 32 = 0.0028$$

2b. Contrast of unknown slopes as

$$\left\{\left[\left(UA_H - UA_L\right)^2\right] \big/ 16\right\} + \left\{\left[\left(UB_H - UB_L\right)^2\right] \big/ 16\right\}$$

– regression squares for unknown

where regression squares for unknown is

$$\left[\left(UA_H + UB_H\right) - \left(UA_L + UB_L\right)\right]^2 \big/ 32$$

thus contrast of unknown slopes is:

$$\left\{\left[(208.1-185.6)^2\right]/16\right\} + \left\{\left[(208.7-185.8)^2\right]/16\right\}$$

$$-\left[(208.1+208.7)-(185.6+185.8)^2\right]/32 = 0.0050$$

The analysis of variance is presented afresh in Table 9.13b.

TABLE 9.13B
Summary of the Analysis of Variance for Example 9.4b, a Two-Dose Level Assay Comparing One Unknown with the Reference Standard Using a Large Plate and an 8 × 8 Latin Square Design[a]

Source of Variance	d.f.	Sum of Squares	Mean Squares	Variance Ratio	Limiting Value	At Arbitrary Probability Level, P	Calculated Probability
Preparations	1	0.316	0.316	10.230	<4.10	0.050	0.003
Regression	1	131.963	131.963	4266.702	>12.60	0.001	<0.001
Parallelism (deviations from)	1	0.019	0.019	0.611	<4.10	0.050	0.439
Contrast of standards	1	0.070	0.070	2.273	<1.70	0.200	0.139
Contrast of unknowns	1	0.020	0.020	0.647	<1.70	0.200	0.425
Contrast of standard slopes	1	0.003	0.003	0.091	<1.70	0.200	0.766
Contrast of unknown slopes	1	0.005	0.005	0.162	<1.70	0.200	0.691
Subtotal	7	132.396					
Treatments	7	132.396					
Rows	7	0.471	0.067	2.176	<2.20	0.050	0.056
Columns	7	0.226	0.032	1.044	<2.20	0.050	0.418
Error by difference	42	1.299	0.031	1.000			
Total	63	134.392					

Note: Limiting values in column 6 refer to the corresponding arbitrary probability level of column 7.

[a] In contrast to Example 9.4a, in this analysis of variance, the contrasts arising from the duplicate weighings are broken down into their component parts to give a more meaningful view of their significance.

Discussion of Assay Validity Based on the Revised Analysis of Variance

The new features of this summary are the values of the two pairs of variance ratios (1) for contrast of standards and contrast of unknowns, which reflect errors in weighing and/or dilutions, and (2) for contrast of standard slopes and contrast of unknown slopes, which reflect errors in dilution only, not in weighings. Thus, the original nomenclature *regression* × *weighings* is erroneous.

None of these variance ratios exceed the limiting value at either the 5% level (4.07) or the 10% level (2.83). If the limiting value at the 20% level (1.70) is applied, we have:

> Variance ratio for contrast of standards (2.273) exceeds the warning limit.
> Variance ratio for contrast of unknowns (0.65) is well within the warning limit.

Thus, in this example even the 20% level test it has been sufficiently sensitive to draw attention to a discrepancy of 5% but not a discrepancy of 3 percent. Before any firm conclusions can be made about the value of this statistical test applied to these two pairs of variance ratios, it would be necessary to look at more assays. However, the evidence presented here suggests that it is of doubtful value.

A FOUR-DOSE LEVEL TURBIDIMETRIC ASSAY FOR ONE UNKNOWN

The assay described now is representative of many performed in a program to develop and validate a turbidimetric assay for gramicidin. It was not the assay design finally chosen because it was found that a three-dose level assay with its narrower overall dose range was more efficient in terms of practical effort and precision attained. However, the four-dose level assay is described here to illustrate the calculation and evaluation procedure.

EXAMPLE 9.5   AN ASSAY OF GRAMICIDIN

Test organism: *Enterococcus durans* NCIB 8192

Design: four replicate tubes for each treatment

Standard: house standard, potency 1069 IU/mg

Dose ratio: 3:2; nominal dose levels: 0.036, 0.054, 0.081, and 0.121 IU/ml

Weighing and dilutions to produce standard test solutions

$$26.2 \text{ mg} \rightarrow 28 \text{ ml} : 0.20 \text{ ml} \rightarrow 50 \text{ ml} \ [4.00 \text{ IU/ml}]$$

Further dilute to give four test solutions thus:

$$440 \ \mu l \rightarrow 50 \ ml \ (0.0352 \ IU/ml)$$

$$660 \ \mu l \rightarrow 50 \ ml \ (0.0528 \ IU/ml)$$

$$990 \ \mu l \rightarrow 50 \ ml \ (0.0792 \ IU/ml)$$

$$1485 \ \mu l \rightarrow 50 \ ml \ (0.1188 \ IU/ml)$$

Unknown: Lozenges stated to contain 200 IU of gramicidin in each lozenge

Dilutions to produce unknown test solutions: disperse 10 lozenges in 500 ml of water (4 IU/ml). Further dilute to give four test solutions in the same manner as for the reference standard.

Responses are recorded as 1000 × optical absorbance at 580 nm in Table 9.14a.

Preliminary Inspection of Data

Referring to Table 9.14A, attention is first focused on ranges of responses and variances within a treatment. The wide ranges and high variances for treatments S_1, U_1, and U_2 were some cause for concern. Inspection suggested that the value 338 in responses to S_1 and 258 in responses to U_2 might be outliers. The *USP* criteria for outliers were applied.

TABLE 9.14A
The Raw Data from Example 9.5, a Four-Dose Level Turbidimetric Assay of Gramicidin, Together with Its Preliminary Inspection

	Observations — Absorbance × 1000 at 580 nm							
	S_1	S_2	S_3	S_4	U_1	U_2	U_3	U_4
	311	244	172	123	284	243	174	113
	338	248	170	112	304	258	173	119
	310	249	176	119	285	238	157	106
	309	247	176	126	300	249	165	110
Treatment totals →	1268	988	694	480	1173	988	669	448
Means →	317	247	174	120	293	247	167	112
Individual slopes →		280	294	214		185	319	221
Overall slopes →				788				725
Range →	29	5	6	14	20	20	17	13
Variances →	196.7	4.7	9.0	36.7	104.9	74.0	62.9	30.0

Criterion 1. For treatment S_1 and U_2 the values of G_1 were obtained by Equation 8.2 as

$$(311 - 338)/(309 - 338) = 0.931$$

and

$$(249 - 258)/(238 - 258) = 0\ 450$$

respectively.

Referring to Appendix 3, part A, it is seen that when $N = 4$, the limiting value for G_1 is 0.846. This does suggest that the value 338 in responses to treatment S_1 is an outlier.

Criterion 2. The sum of the ranges is

$$29 + 5 + 6 + 14 + 20 + 20 + 17 + 13 = 124$$

R_* was obtained as

$$\text{largest range/sum of ranges} = 29/124 = 0.234$$

Referring to Appendix 3, part B, the limiting value for R_* for eight ranges each including four observations is 0.267. As the calculated value (0.234) is less than this limiting value, this suggests that the observations do not include an outlier.

To assess curvature and parallelism the mean responses to each treatment were plotted against log dose (Figure 9.4) and joined point to point. The line for the unknown exhibited the expected slightly sigmoid character, whereas the line for the standard was straight between response to doses S_1 to S_3. On omitting the response 338 to dose S_1, some sigmoid character was introduced, which is expected in this assay.

From consideration of all the evidence, the judgment was reached that it would be safe to reject the response 338 to dose S_1. Table 9.14b shows the responses after replacement of the value 338 by the mean of other responses to the same treatment, 310. Referring now to Table 9.14b:

1. Slopes expressed as the difference in treatment totals, high dose – low dose, for both standard and unknown at 760 and 725 appear close enough and probably indicate parallel responses within the expected error.
2. The sigmoid curvature that has already been noted will not necessarily invalidate the assay. Later, the analysis of variance will reveal whether this curvature can be attributed mainly to a quadratic component; if this is shown to be the case, then there would be no reason to reject the assay on this account.
3. Inspection of the values for variance within each treatment group does not suggest any correlation between the magnitude of variance and dose level. Thus, the assay accords with assumption 4 set forth at the beginning of Chapter 8.

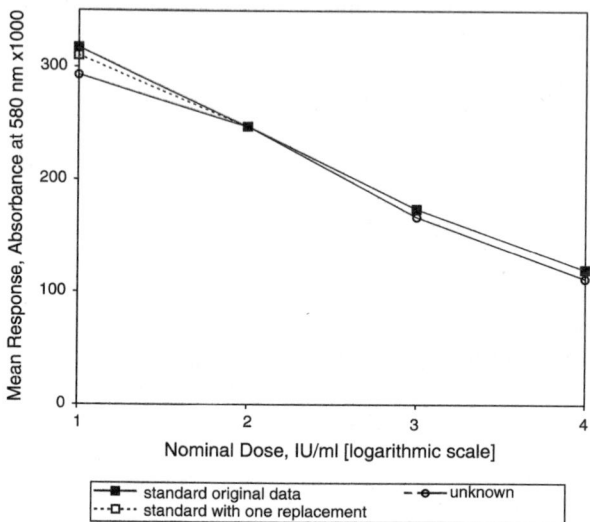

FIGURE 9.4 A graphical representation of the raw data from Example 9.5, A Four-Dose Level Turbidimetric Assay of Gramicidin. Mean responses to each test solution are plotted. Continuous lines represent the original data; the dotted line indicates the change when one outlier has been replaced.

TABLE 9.14B
The Raw Data from Example 9.5, with One Replacement Value (in Bold Italics), a Four-Dose Level Turbidimetric Assay of Gramicidin Together with a Further Preliminary Inspection

	Observations — Absorbance × 1000 at 580 nm							
	S_1	S_2	S_3	S_4	U_1	U_2	U_3	U_4
	311	244	172	123	284	243	174	113
	310	248	170	112	304	258	173	119
	310	249	176	119	285	238	157	106
	309	247	176	126	300	249	165	110
Treatment totals →	1240	988	694	480	1173	988	669	448
Means →	310	247	174	120	293	247	167	112
Individual slopes →		252	294	214		185	319	221
Overall slopes →				760				725
Range →	2	5	6	14	20	20	17	13
Variances →	0.7	4.7	9.0	36.7	104.9	74.0	62.9	30.0

In consideration of all the evidence from inspection of these data including one replacement value, it was deemed appropriate to go ahead with the calculation of potency estimate and standard statistical evaluation.

Calculation of Potency Estimate from the Treatment Totals *

Step 1. Calculate E, the weighted mean increase in response for a 1.5-fold increase in dose on the basis of the expression given in Appendix 1. The rationale for this expression is illustrated by means of Figure 9.5, in which the greater weight apportioned to doses 1 and 4 is analogous to the greater moments of forces applied to a lever at greater distances from the fulcrum. Arbitrarily setting the mean log dose at zero and fitting coded log doses so that log 1.5 is coded as 2, it is seen that the deviations from the mean coded log dose are

$$S_1 \text{ and } T_1 = -3, \quad S_2 \text{ and } T_2 = -1,$$

$$S_3 \text{ and } T_3 = +1, \quad S_4 \text{ and } T_4 = +3$$

Applying weighting factors of 3 to the extreme dose levels and 1 to the intermediate dose levels, the total of weighted *distances* from the center point (zero) for each preparation is

$$(3 \times 3) + (1 \times 1) + (1 \times 1) + (3 \times 3) = 20$$

and so for both preparations is 40. That is, the grand total of weighted distances from the center point is $40 \times (0.5 \times \log 1.5)$, or $20 \times \log 1.5$. Hence, the value 20 in the denominator so that the expression corresponds to one log dose interval.

Because treatment totals are used instead of treatment means and the replication is four, the denominator is changed from 20 to (20×4)

Dose Level Number	1		2	midpoint	3		4
				<------------------->			
		<-->					
Coded Log Dose	-3		-1	0	+1		+3

FIGURE 9.5 An illustration of the rationale in the weighting of responses. The greater weight apportioned to doses 1 and 4 is analogous to the greater moments of forces applied to a lever at increasing distances from the fulcrum..

$$E = \left[3\left(S_4 + T_4\right) + \left(S_3 + T_3\right) - \left(S_2 + T_2\right) - 3\left(S_1 + T_1\right)\right]/(20 \times 4)$$

$$= \left[3(480 + 448) + (694 + 669) - (988 + 988) - 3(1268 + 1173)\right]/80$$

$$= -63.35$$

Step 2. Calculate F, the mean difference between unknown and standard responses.

$$F = \left[\left(T_4 + T_3 + T_2 + T_1\right) - \left(S_4 + S_3 + S_2 + S_1\right)\right]/(4 \times 4)$$

As in the case of E, because treatment totals are used instead of treatment means and the replication is four, the denominator is changed from 4 to (4 × 4), so that

$$F = \left[(448 + 669 + 988 + 1173) - (480 + 694 + 988 + 1240)\right]/16$$

$$= -7.75$$

Step 3. Calculate b, the mean difference in response for a tenfold increase in dose.

$$b = E/I = -63.35/\log 1.5 = -359.8$$

Step 4. Calculate M, the logarithm of the ratio of the potency of unknown to that of the standard.

$$M = F/b = -7.75/-359.8 = 0.02154$$

Step 5. Calculate potency estimate of high-dose test solution of unknown relative to that of the standard high-dose test solution, then of the unknown itself.

Relative potency estimate of high-dose test solution of unknown is

$$\text{antilog } M = 1.0509$$

As potency of the standard test solution is 0.1188 IU/ml, the estimated potency of the unknown test solution is

$$1.0509 \times 0.1188 \text{ IU/ml} = 0.1248 \text{ IU/ml}$$

so that the estimated potency of a lozenge is given by

$$\frac{0.1248 \times 50 \times 1000 \times 500}{1484 \times 10} = 210 \text{ IU/lozenge}$$

Statistical Evaluation

Step 1. From the raw data of Table 9.14B (including the replacement value), calculate the total deviation squares as

$$311^2 + 310^2 + \ldots + 106^2 + 110^2 - 6680^2/32$$

$$= 1,557,498 - 1,394,450 = 163,048$$

Step 2. Using orthogonal polynomial coefficients (OPCs), calculate the deviation squares attributable to:

Preparations
Regression
Parallelism
Quadratic curvature
Opposed quadratic curvature
Cubic curvature
Opposed cubic curvature

The calculation is shown in tabular form in Table 9.15. The figures in columns 10 to 12 are calculated thus:

Column 10, headed e_i, each figure is the sum of the squares of the eight OPCs in the same row;
Column 11, headed T_i, each figure is the sum of the products of the individual treatment totals (bottom row) and the corresponding OPC (in the same column);
Column 12, the heading $T_i^2/4e_i$ is self-explanatory.

Step 3. Calculate the treatment deviation squares as

$$\frac{\left(1240^2 + 988^2 + 694^2 + 480^2 + 1173^2 + 988^2 + 669^2 + 448^2\right)}{4} - \frac{6680^2}{32}$$

$$= 1,556,529.5 - 1,394,450 = 162,079.5$$

Step 4. Summarize the analysis of variance in Table 9.16. It will be noted that the sum of all the components calculated in step 2 is almost identical with the treatment deviation squares calculated in step 3. (Failure to be identical or almost identical would indicate a computational error).

TABLE 9.15
Factorial Coefficients Applied to the Data of Example 9.5, a Four-Dose Level Symmetrical Turbidimetric Assay

Row	S_1	S_2	S_3	S_4	U_1	U_2	U_3	U_4	e_i	T_i	$T_i^2/(4 \times e_i)$
a	−1	−1	−1	−1	1	1	1	1	8	−124	480.50
b	−3	−1	1	3	−3	−1	1	3	40	−5068	160,528.90
ab	−3	−1	1	3	3	1	−1	−3	40	−80	40.00
q	1	−1	−1	1	1	−1	−1	1	8	2	0.13
aq	1	−1	−1	1	−1	1	1	−1	8	74	171.13
c	−1	3	−3	1	−1	3	−3	1	40	354	783.23
ac	−1	3	−3	1	1	−3	3	−1	40	−110	75.63
Treatment totals →	1240	988	694	480	1173	988	669	448			162,079.50

TABLE 9.16
Summary of Analysis of Variance, Example 9.5, a Four-Dose Level Turbidimetric Assay Using a 3:2 Dose Ratio

Source of Variance	d.f.	Sum of Squares	Mean Squares	Variance Ratio	Limiting Value	At Arbitrary Probability Level, P	Calculated Probability
Preparations	1	480.5	480.5	11.41	<4.28	0.050	0.003
Regression	1	160,528.9	160,528.9	3,811.86	>14.19	0.001	<0.001
Parallelism (deviations from)	1	40.0	40.0	0.95	<4.28	0.050	0.340
Quadratic curvature	1	0.1	0.1	0.00	<4.28	0.050	0.961
Opposed quad. Curvature	1	171.1	171.1	4.06	<4.28	0.050	0.056
Cubic curvature	1	783.2	783.2	18.60	<4.28	0.050	0.000
Opposed cubic curvature	1	75.6	75.6	1.80	<4.28	0.050	0.193
Subtotal	7	162,079.4					
Treatments	7	162,079.5					
Error by difference	23	968.6	42.1	1.00			
Total	30	163,048.0					

Note: Limiting values in column 6 refer to the corresponding arbitrary probability level of column 7.

In Table 9.16, error mean squares is obtained as the difference between total deviation squares and treatment deviation squares. It will be noted that the total degrees of freedom and degrees of freedom for residual error are 30 and 23, respectively, rather than 31 and 24. This is because of the use of one replacement value.

Mean squares and variance ratios are calculated in the usual manner.

Discussion of Assay Validity Based on the Analysis of Variance

In this assay, the pharmacopoeial requirements concerning variance ratios apply to regression, parallelism, and nonlinearity. The latter has been broken down into its component parts: quadratic curvature, opposed quadratic curvature, cubic curvature, and opposed cubic curvature.

Variance ratio for regression at almost 4000 is, typically, highly satisfactory. That for parallelism at 0.95 is well below the limit of 4.28. Of the variance ratios for the four components of nonlinearity, only that for cubic curvature at 18.6 exceeds the limiting value of 4.28.

Variance ratio for preparations at 11.41 is well above the warning level of 4.28. However, the estimated potency of the unknown differed from that of the reference standard by only 5.1%. Such a small difference should not be cause for concern about the validity of the assay. However, the value of the test itself may be questioned.

In conclusion, with a variance ratio of 18.6, cubic curvature is highly significant as foreseen in the preliminary inspection. Although this throws some doubt on the validity of the assay, the parallel-line calculation is very robust and was continued as a matter of interest and for comparison with a possible alternative calculation using only three dose levels.

Step 5. Calculate the term S_{xx}, the sum of the squares of deviations of individual log doses from the mean log dose. The dose ratio is 3:2, so the log dose interval is 0.1761.

Setting the mid-dose level at 1 (log mid dose = 0), log intervals from the mid dose are

$$-1.5 \times 0.1761, \; -0.5 \times 0.1761, \; +0.5 \times 0.1761 \text{ and } +1.5 \times 0.1761$$

There are two preparations and the replication is four, so the value of S_{xx} is obtained as

$$S_{xx} = 2 \times 4 \times [(-0.26414)^2 + (-0.08805)^2 + (-0.08805)^2 + (-0.26414)^2] = 1.24033$$

Calculate g, the index of significance of the slope, b, using Equation (9.22), as

$$g = \frac{42.11 \times 2.069^2}{-359.8^2 \times 1.24033} = 0.001123$$

where s^2, b, and S_{xx} have the values already calculated, and t is Student's t obtained from tables for $P = 0.95$ and 23 degrees of freedom as 2.069.

As g is below 0.1, calculate confidence limits of the potency estimates by the approximate method.

First calculate $V(M)$, the variance of the logarithm of the potency estimate of the unknown, by substituting the known values in the Equation (9.23)

$$V(M) = \frac{42.11}{-359.8^2}\left[\frac{1}{16} + \frac{1}{16} + \frac{0.02154^2}{1.24033}\right]$$

$$= 0.0003253[0.0625 + 0.0625 + 0.000374] = 0.000040784$$

The standard error of the logarithm of the potency estimate, s_m, is then obtained by Equation (9.24) as

$$s_m = (0.000040784)^{0.5} = 0.006386$$

Log percent confidence limits are given by Equation (9.25) as

$$2 \pm 2.069 \times 0.006386 = 1.9868 \text{ and } 2.0132$$

The corresponding percentage limits are 97.0 and 103.1% ($P = 0.95$).

Alternatively, expressed in terms of the standard potency, the estimated potency is 105.1% with limits 101.9 to 108.3%, or 204 to 217 IU/lozenge ($P = 0.95$).

A REVISED CALCULATION USING ONLY THREE-DOSE LEVELS

In view of the significant cubic curvature revealed by the analysis of variance, the estimated potency calculation and evaluation were repeated using only three dose levels. Because the responses to treatment S_1 do not lie on the expected somewhat sigmoid line it was decided to recalculate using dose levels two, three, and four.

Step 1. Calculate E, the mean increase in response for a 1.5-fold increase in dose, on the basis of the expression given in Appendix 1. Because treatment totals are used instead of treatment means and the replication is four, the denominator is changed from 4 to (4 × 4)

$$E = \left[(S_4 + T_4) - (S_2 + T_2)\right]/(4 \times 4)$$

$$= \left[(480 + 448) - (988 + 988)\right]/(4 \times 4)$$

$$= -65.50$$

Step 2. Calculate F, the mean difference between unknown and standard responses.

As in the case of E, because treatment totals are used instead of treatment means and the replication is four, the denominator is changed from 3 to (3×4) so that

$$F = \left[\left(T_4 + T_3 + T_2\right) - \left(S_4 + S_3 + S_2\right)\right]/(3 \times 4)$$

Putting in the treatment totals

$$F = \left[(448 + 669 + 988) - (480 + 694 + 988)\right]/(3 \times 4)$$

$$= -4.75$$

Step 3. Calculate b, the mean difference in response for a tenfold increase in dose.

$$b = E/I = -65.50/\log 1.5 = -372.5$$

Step 4. Calculate M, the logarithm of the ratio of the potency of unknown to that of the standard.

$$M = F/b = -4.75/-372.5 = 0.01275$$

Step 5. Calculate potency estimate of high-dose test solution of unknown relative to that of the standard high-dose test solution, then of the unknown itself.

As potency of the standard test solution is 0.1188 IU/ml, the estimated potency of the unknown test solution is:

$$1.0298 \times 0.1188 \text{ IU/ml} = 0.1223 \text{ IU/ml}$$

so that the estimated potency of a lozenge is given by:

$$\frac{0.1223 \times 50 \times 1000 \times 500}{1484 \times 10} = 206 \text{ IU/lozenge}$$

Relative potency estimate of high-dose test solution of unknown is

$$\text{antilog } M = 1.0298$$

Statistical Evaluation

Step 1. Using a part of the modified raw data of Table 9.14B calculate the total deviation squares as

$$244^2 + 248^2 + \ldots + 106^2 + 110^2 - 4267^2/24$$

$$= 828,799 - 758,637 = 70,162$$

Step 2. Using OPCs calculate the deviation squares attributable to:

Source	Row
Preparations	a
Regression	b
Parallelism	ab
Quadratic curvature	q
Opposed quadratic curvature	aq

This calculation is shown in Table 9.17.

Step 3. Calculate the treatment deviation squares as

$$\frac{\left(988^2 + 694^2 + 480^2 + 988^2 + 669^2 + 448^2\right)}{4} - \frac{4267^2}{24}$$

$$= 828,147.3 - 758,637 = 69,510.21$$

Step 4. Summarize the analysis of variance in Table 9.18. It will be noted that the sum of all the components calculated in step 2 is almost identical with the treatment deviation squares calculated in step 3. (Failure to be identical or almost identical would indicate a computational error).

In Table 9.18, error mean squares is obtained as the difference between total deviation squares and treatment deviation squares. Mean squares and variance ratios are calculated in the usual manner.

Discussion of Assay Validity Based on the Revised Analysis of Variance

The analysis of variance summarized in Table 9.18 is compared with that of Table 9.16. The variance ratio for regression remains highly significant; that for parallelism remains nonsignificant. The undoubted curvature of the response lines is now revealed as quadratic curvature (because cubic curvature cannot be detected using only three points on the curve). With a variance ratio of 18.23, quadratic curvature is very significant. However, regardless of pharmacopoeial concerns, a component of quadratic curvature does not invalidate a symmetrical parallel-line assay (see Chapter 11). Variance ratio for opposed quadratic curvature is not significant.

The value for preparations is now below the warning limit, but it is of little consequence whether it is below or above the limit. Thus, on the basis of this analysis of variance, there is no reason to doubt the validity of the assay,

Step 5. Calculate the term S_{xx}, the sum of the squares of deviations of individual log doses from the mean log dose. The dose ratio is 3:2, so the log dose interval is 0.1761. Setting the mid-dose level at 1 (log mid dose = 0),

TABLE 9.17
Factorial Coefficients Applied to Part of the Data of Example 9.5, a Four-Dose Level Symmetrical Turbidimetric Assay as Part of the Analysis of Variance[a]

Row	S_2	S_3	S_4	U_2	U_3	U_4	e_i	T_i	$T_i^2/(4 \times e_i)$
a	−1	−1	−1	1	1	1	6	−57	135.38
b	−1	0	1	−1	0	1	4	−1048	68,644.00
ab	1	0	−1	−1	0	1	4	−32	64.00
q	1	−2	1	1	−2	1	12	178	660.08
aq	−1	2	−1	1	−2	1	12	18	6.75
Treatment totals →	988	694	480	988	669	448			

[a] In this calculation, responses to dose-level 1 have been discarded so as to select a part of the response line more closely approaching a straight line.

TABLE 9.18
Summary of Analysis of Variance, Example 9.5, a Turbidimetric Assay Using a 3:2 Dose Ratio

Source of Variance	d.f.	Sum of Squares	Mean Squares	Variance Ratio	Limiting Value	At Arbitrary Probability Level, P	Calculated Probability
Preparations	1	135.40	135.40	3.740	<4.41	0.050	0.069
Regression	1	68,644.00	68,644.00	1,895.952	>15.38	0.001	<0.001
Parallelism (deviations from)	1	64.00	64.00	1.768	<4.41	0.050	0.200
Quadratic curvature	1	660.10	660.10	18.232	<4.41	0.050	0.001
Opposed curvature	1	6.80	6.80	0.188	<4.41	0,050	0.668
Subtotal	5	69,510.30					
Treatments	5	69,510.30					
Error by difference	18	651.70	36.21	1.000			
Total	23	70,162.00					

Note: Limiting values in column 6 refer to the corresponding arbitrary probability level of column 7.

Note: This calculation is based on only three of the four dose levels.

log intervals from the mid dose are 0.1761. The replication is four for each of the two preparations, so the value of S_{xx} is obtained as

$$S_{xx} = 2 \times 4 \times [(0.1761)^2 + (-0.1761)^2] = 0.49613$$

Step 6. Calculate g, the index of significance of the slope, b, using Equation (9.22)

$$g = \frac{36.2 \times 2.101^2}{-372.0^2 \times 0.49613} = 0.000573$$

where s^2, b, and S_{xx} have the values already calculated, and t is Student's t obtained from tables for $P = 0.95$ and 18 degrees of freedom as 2.101.

Because g is below 0.1, calculate confidence limits of the potency estimates by the approximate method.

Step 7. Calculate $V(M)$, the variance of the logarithm of the potency estimate of the unknown, by substituting the known values in Equation (9.23)

$$V(M) = \frac{36.2}{-372.0^2}\left[\frac{1}{12} + \frac{1}{12} + \frac{0.01277^2}{0.49613}\right] =$$

$$0.0002616[0.083333 + 0.083333 + 0.000329] = 0.000043684$$

Step 8. The standard error of the logarithm of the potency estimate, s_m, is then obtained by Equation (9.24) as

$$s_m = (0.000043684)^{0.5} = 0.006609$$

Step 9. Log percent confidence limits are given by Equation (9.25) as

$$2 \pm 2.101 \times 0.006609 = 1.9861 \text{ and } 2.0139$$

Step 10. The corresponding percentage limits are 96.9 and 103.2% ($P = 0.95$).

The results of the three-dose level and four-dose level calculations are summarized in tabular form:

	Three-Dose Level Calculation, %	Four-Dose Level Calculation, %
Potency relative to reference standard	103.0	105.1
Confidence limits as percent of estimated potency ($P = 0.95$)		
Lower confidence limit	96.9	97.0
Upper confidence limit	103.2	103.1

The potency estimate based on three dose levels is the preferred result for two reasons:

1. The validity of the estimate based on four dose levels is suspect due to the significant component of cubic curvature.
2. The mean response to test solution S_1 does not lie on the expected some-what sigmoid curve and may include some aberrant responses.

A SMALL PLATE ASSAY USING A FIVE-DOSE LEVEL STANDARD CURVE

The term *5 + 1 assay* is used here as the title for an assay design that is described in the *United States Pharmacopoeia* and is the preferred method in the U.S. It employs a five-dose level standard curve and a single dose of each unknown preparation. It is classified here as a parallel-line assay because the assumption of calculation procedures is that response is directly proportional to logarithm of dose; this is the basis of parallel-line assays.

The ratio between adjacent dose levels for the plate assay is almost invariably 5:4, which gives an overall dose range of about 2.44:1 — a reasonably narrow range. (The 5 + 1 design is also applied to turbidimetric assays, and in those cases, the dose ratio is often narrower, e.g., 9:8).

Unlike most other plate assay designs that have been described, each plate includes only two treatments — the reference treatment (mid-dose standard) and either one of the other four-dose levels of the reference standard or one of the unknowns. Thus, each plate does not represent a self-contained assay. Because of interplate variation, the different treatments are not directly comparable but are compared with a common reference point — the mean response to reference dose on those plates representing the standard curve. A correction is applied to the responses to standards; then the corrected values are used to define the standard curve.

Although it is normal to compare many unknowns with the standard in the same assay, the example given here shows only one unknown.

EXAMPLE **9.6**   AN ASSAY OF KANAMYCIN

Test organism: *Bacillus pumilis* NCTC 8241

Design: US CFR petri dish assay

Standard: house standard, potency 812 IU/mg

Dose ratio: 5:4

Nominal dose levels of standard 3.2, 4.0, 5.0 (reference point), 6.25, and 7.81 IU/ml

Weighing and dilution for standard test solutions

$$31.5 \text{ mg} \rightarrow 25.6 \text{ ml} : 2 \text{ ml} \rightarrow 200 \text{ ml}$$

Giving a master standard solution of potency

$$\frac{31.5 \times 812 \times 2}{25.6 \times 200} = 9.991 \text{ IU/ml}$$

Nominal potency of this solution is 10.00 IU/ml; therefore, a factor of 0.9991 is applied in the calculation of sample potency estimate.

Further dilutions of the standard were:

$$3.2 \text{ ml} \rightarrow 10 \text{ ml} \quad \text{standard 1} \quad 3.2 \text{ IU/ml}$$

$$4.0 \text{ ml} \rightarrow 10 \text{ ml} \quad \text{standard 2} \quad 4.0 \text{ IU/ml}$$

$$5.0 \text{ ml} \rightarrow 10 \text{ ml} \quad \text{reference} \quad 4.0 \text{ IU/ml}$$

$$6.25 \text{ ml} \rightarrow 10 \text{ ml} \quad \text{standard 4} \quad 6.25 \text{ IU/ml}$$

$$7.81 \text{ ml} \rightarrow 10 \text{ ml} \quad \text{standard 5} \quad 7.81 \text{ IU/ml}$$

Sample: Kanamycin injection of labeled potency 1 g kanamycin base/vial.

Mean weight of contents of 10 vials was 1.284 g.

Weighing and dilution to unknown test solution

$$246.1 \text{ mg} \rightarrow 200 \text{ ml} : 5 \text{ ml} \rightarrow 100 \text{ ml} : 10 \text{ ml} \rightarrow 100 \text{ ml}$$

Giving a concentration of

$$\frac{246.1 \times 1000 \times 5 \times 10}{200 \times 100 \times 100} = 6.1525 \text{ µg/ml}$$

The pattern of distribution of test solutions on plates is shown in Appendix 7.

The correction procedure is illustrated by Table 9.19, which presents the raw data. Each block of six figures in the table represents the zone diameters of a single plate. Taking the top left block, the responses to low-dose standard are 1460, 1410, and 1380; responses to the reference dose (mid-dose standard) are 1610, 1560, and 1580. These figures are zone diameters in mm × 100. The mean for all nine reference zones of that group of three plates (that has the low-dose standard) is shown in the penultimate row of the table to be 1586.7. Similarly the means of the reference zones for the other three groups corresponding to the standards are 1556.7, 1578.9, and 1566.7. The grand mean of these four is 1572.2. It is then seen that the deviations of the groups of three means from the grand mean are

$$1586.7 - 1572.2 = +14.5; \quad 1556.7 - 1572.2 = -15.5$$

$$1578.9 - 1572.2 = +6.7; \quad 1566.7 - 1572.2 = -5.5$$

These were rounded off by the computer to +14, −16, +7, and −6.

TABLE 9.19
The Uncorrected Data for Example 9.6, an Assay of Kanamycin

Zone Diameters (mm × 100)

S_1	R	S_2	R	S_4	R	S_5	R	U	R
1460	1610	1470	1580	1660	1560	1730	1560	1530	1570
1410	1560	1510	1560	1680	1580	1700	1560	1580	1580
1380	1580	1480	1550	1630	1600	1700	1550	1570	1570
1450	1600	1470	1570	1660	1580	1730	1560	1580	1590
1410	1590	1490	1550	1650	1560	1740	1570	1580	1570
1440	1620	1520	1560	1620	1570	1720	1550	1550	1570
1400	1570	1480	1570	1690	1610	1730	1590	1520	1550
1420	1570	1500	1540	1650	1570	1730	1580	1510	1580
1410	1580	1430	1530	1680	1580	1670	1580	1510	1530
Treatment totals → 12,780	14,280	13,350	14,010	14,920	14,210	15,450	14,100	13,930	14,110
Means for reference →	1586.7		1556.7		1578.9		1566.7		1567.8
Means for nonreference → 1420.0		1483.3		1657.8		1716.7		1547.8	
Corrections →		14		−16		7		−6	

Grand mean of *four* references 1572.2

It is inferred from these figures that, on average, all zones in the first set of three plates are each too big by about 14 and so corrected zone sizes for the standards 1460 are calculated as

$$1460 - 14 = 1446$$

and so on.

The corrected responses to standard dose levels 1, 2, 4, and 5 are shown in Table 9.20. In the routine calculation of potency estimate for the unknown, it is not necessary to calculate corrected responses for the unknown because potency estimate is based on a direct comparison between the mean for unknown and mean for the reference responses in the set of three plates. However, to carry out an analysis of variance, it is necessary that all responses be directly comparable, and so corrected responses for the unknown are given in the last column of Table 9.20.

Preliminary Inspection of Data

Considering the data of Table 9.20:

1. Ranges of corrected responses within treatments are somewhat wide (minimum 70, maximum 110, corresponding to 0.7 to 1.1 mm) but not sufficiently wide to cause undue concern.
2. Variances range from 500 to 1050; there is no reason to suspect that size of variance is related to dose.
3. There is some overlap of ranges; thus there is a value 1446 in both columns S_1 and S_2; there is a 1683 in column S_4 and the slightly lower value 1676 in column S_5. However, bear in mind that the ratio between adjacent dose levels is only 5:4 and so these overlaps seem acceptable considering the normal random variation to be expected in the agar diffusion assay.
4. Individual slopes expressed as differences between totals of responses for adjacent dose levels are reasonably consistent and perhaps show some curvature in the expected direction. Again bear in mind the narrow ratio between adjacent dose levels; less consistent slopes than these would still not be cause for alarm.
5. Both *USP* tests for outliers suggest that all is well.

Calculation of the Potency Estimate from the Treatment Total

On the basis of this preliminary inspection, there is no reason to doubt the soundness of the raw data, and so it is appropriate to proceed with the routine calculation of potency estimate and statistical evaluation.

As this assay is by a U.S. "Code of Federal Regulations" design, the U.S. calculation guidance is described first. There are two alternative graphical methods:

1. Plot the mean corrected responses to standard (and midpoint reference) on log paper against logarithm of dose and draw a line through these points by inspection or

TABLE 9.20
The Corrected Data of Example 9.6 Together with Preliminary Inspection

	S_1	S_2	R	S_4	S_5	U
					Zone Diameters (mm × 100)	
	1446	1486	1572	1653	1736	1534
	1396	1526	1572	1673	1706	1584
	1336	1496	1572	1623	1706	1574
	1436	1486	1572	1653	1736	1584
	1396	1506	1572	1643	1746	1584
	1426	1536	1572	1613	1726	1554
	1386	1496	1572	1683	1736	1524
	1406	1516	1572	1643	1736	1514
	1396	1446	1572	1673	1676	1514
Treatment totals →	12,624	13,494	14,148	14,857	15,504	13,966
Means →	1403	1499	1572	1651	1723	1552
Ranges →	110	90	NA	70	70	70
Variances →	1050	700	NA	544	500	944
Slope inspection →		870	654	709	647	
USP criterion 1						
G2a→	0.600	0.500	NA	0.500	0.500	0.143
G2b →	0.333	0.400	NA	0.167	0.250	0.000

Limiting value of G_2 for 9 responses is 0.780, so on this basis there is no reason to suspect an outlier.

USP criterion 2
R*→   0.268

Limiting value of R* for 9 responses in each of 5 groups is 0.320, so on this basis there is no reason to suspect an outlier.

NA means non applicable.

2. From the corrected responses to standard calculate (best estimate) the response to low and high doses (L and H) from the following equations:

$$L = (3a + 2b + c - e)/5 \qquad (9.26)$$

$$H = (3e - 2d + c - a)/5 \qquad (9.27)$$

where

c = average zone diameter of the 36 readings of response to reference point (mid-dose) of standard

a, b, d, e = corrected average values for the other standard solutions, lowest to highest concentrations, respectively.

Draw a straight line joining L and H.

To estimate the potency of the unknown sample (whichever way the graph has been obtained) for the three plates representing the unknown, calculate the mean responses for both sample and reference points and obtain their difference. If the average zone diameter of the sample is larger than that of the reference, add that difference to the reference concentration diameter of the standard response line (grand mean of reference points) and read the corresponding concentration from the response line. Similarly, if the average zone diameter of the sample is smaller than that of the reference, subtract that difference from the reference concentration diameter.

The graphical procedure is illustrated in Figure 9.6, in which the graph was drawn using the second procedure.

Thus using Equation (9.26) and Equation (9.27):

$$L = \left[(3 \times 1403) + (2 \times 1499) + 1572 - 1723 \right] / 5 = 1411.2 \quad (14.11 \text{ mm})$$

$$H = \left[(3 \times 1723) + (2 \times 1651) + 1572 - 1403 \right] / 5 = 1728.0 \quad (17.28 \text{ mm})$$

From the graph, the mean corrected zone diameter of the unknown (15.48 mm) corresponded to −0.023 on the log scale relative to the reference point. Antilog −0.023 is 0.948. Thus, the graphical procedure shows the unknown to have a potency of 94.8% of that of the reference solution.

The use of an arithmetical procedure for interpolation from the graph seems rather pointless after the imprecision of drawing a graph has been introduced. Bearing in mind that any curvature of the response line over such a short overall dose range (1:2.44) would be negligible, it is better to use a purely arithmetical procedure that assumes a straight line. The potency estimate may be calculated by the purely arithmetical procedures that have been used in other examples, calculating E and F from the expressions in the tabulation of Appendix 1, which are related to assay design and replication.

Step 1. Calculate E, the mean increase in response for a 1.25-fold increase in dose, using only the corrected response totals to the four levels of standard plus the reference (midpoint standard).

$$E = [(2 \times 15,505) + 14,857 - 13,494 - (2 \times 12,624)]/(10 \times 9) = 79.1667$$

Step 2. Calculate F, the mean difference in (uncorrected) responses between the unknown and the reference for the final group of three plates.

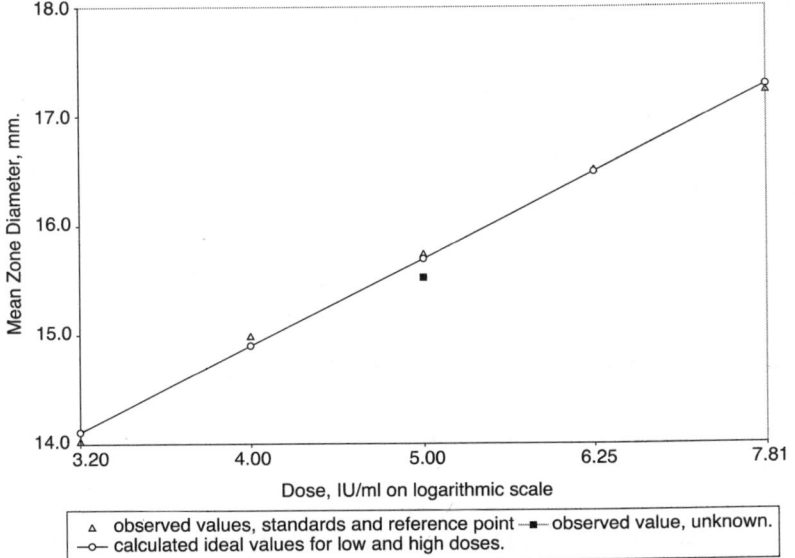

FIGURE 9.6 A graphical representation of the data of example 9.6. It is based on the corrected mean zone sizes of Table 9.20. The straight line is drawn to join the values of L and H that were calculated using Equation (9.26) and Equation (9.27), respectively.

$$F = (13{,}930 - 14{,}110)/9 = -20$$

Step 3. Calculate b, the mean difference in response for a tenfold increase in dose.

$$b = E/I = 79.1667/\log_{10} 1.25 = 816.909$$

Step 4. Calculate the value of M, the logarithm of the ratio of the potency of unknown to that of the standard.

$$M = F/b = -20/816.909 = -0.0244825$$

Step 5. Calculate potency estimate of the unknown test solution relative to that of the reference test solution.

$$\text{antilog}_{10} M = 0.9452$$

(Compare this with 0.948 obtained graphically.)

Step 6. Calculate potency estimate of the sample. Actual potency estimate of the unknown test solution is

$$0.9452 \times 0.9991 \times 5 = 4.722 \text{ IU/ml}$$

The concentration of the test solution is 6.1525 µg/ml (or 0.0061525 mg/ml) so that estimated potency is

$$4.722/0.0061525 = 767.49 \text{ IU/mg}$$

Therefore, a vial having an average weight of contents of 1.284 g is estimated to contain

$$\frac{767.49 \times 1284}{1000} = 985 \text{ mg of kanamycin base}$$

Statistical Evaluation

Step 1. From the corrected data of Table 9.20, calculate the total deviation squares as

$$\left[1446^2 + 1486^2 + 1572^2 + \ldots + 1676 + 1514^2\right] - 84,593^2/54$$

$$= 113,115,793.00 - 132,518,067.57 = 597,725.43$$

$$\left[\text{mean correction, } MC = 84,593^2/54 = 132,518,067.57\right]$$

Step 2. From the treatment totals, calculate

$$\left[\left(12,624^2 + 13,494^2 + 14,148^2 + 14,857^2 + 15,504^2 + 13,966^2\right)/9\right] - MC$$

$$= 133,085,881,329.90 - 132,518,067.57 = 567,814.31$$

Step 3. From the preparation totals, calculate

$$\left[\left(12,624 + 13,494 + 14,148 + 14,857 + 15,504\right)^2/45\right] + \left(13,966^2\right)/9 - MC$$

$$= 110,848,291.80 + 21,672,128.44 - 132,518,067.57 = 2352.63$$

Step 4. Using orthogonal polynomial coefficients, calculate squares for regression and deviations from regression squares as shown in Table 9.21.

Step 5. Insert the known deviation squares in Table 9.22, the summary of analysis of variance, and calculate the residual error squares by difference.

Discussion of Assay Validity Based on the Analysis of Variance

First, a comment on the number of degrees of freedom is appropriate. For the calculations using orthogonal polynomial coefficients, the value of the reference

TABLE 9.21
Factorial Coefficients for a Five-Dose Level Standard Curve Applied to the Data of Example 9.10

	S_1	S_2	S_3	S_4	S_5	e_i	T_i	$T_i^2/(9 \times e_i)$
b	−2	−1	0	1	2	10	7123	563,745.88
q	2	−1	−2	−1	2	14	−391	1213.34
r	−1	2	0	−2	1	10	154	263.51
s	1	−4	6	−4	1	70	−388	238.96
Treatment totals →	12,624	13,494	14,148	14,857	15,504			565,461.69

TABLE 9.22
Summary of Analysis of Variance for Example 9.6, an Assay Using Small Plates and the *USP* Five-Dose Level Standard Curve Procedure

Source of Variance	d.f.	Sum of Squares	Mean Squares	Variance Ratio	Limiting Value	At Arbitrary Probability Level, P	Calculated Probability
Preparations	1	2,352.6	2,352.6	3.150	<4.35	0.050	0.084
Regression	1	563,745.9	563,745.9	753.895	>12.61	0.001	<0.001
Quadratic curvature	1	1,213.3	1,213.3	1.623	<4.35	0.050	0.210
Cubic curvature	1	263.5	263.5	0.320	<4.35	0.050	0.575
Quartic curvature	1	239.0	239.0	0.320	<4.35	0.050	0.575
Subtotal	5	567,814.3					
Treatments	5	567,814.3	113,562.9				
Error by difference	40	29,911.1	747.8	1.000			
Total	45	597,725.4					

Note: Limiting values in column 6 refer to the corresponding arbitrary probability level of column 7.

Note: Adjacent dose levels of standard were in the ratio 5:4. There was a single level of the unknown preparation.

point was used effectively nine times, so that the number of responses used in the calculation was 54, which would normally lead to a total of 53 degrees of freedom. However, there were not nine *independent* reference points, and so effectively there were only 46 independent responses leading to a total of 45 degrees of freedom.

In this assay, the pharmacopoeial requirements concerning variance ratios apply to responses to the five standard treatments only, that is, to regression and deviations from linearity. The latter is broken down into its component parts: quadratic, cubic, and quartic curvature. Variance ratios for the three deviations from linear regression were all well within the limit and not statistically significant. However, the reliability of these figures is questionable. The preliminary inspection of data showed that the ranges of responses to adjacent treatment levels overlapped in two cases. This overlap suggested, therefore, that the figures for cubic and quartic curvature may well be due to random error. Although quadratic curvature might be expected on theoretical grounds, this assay with an overall short dose range may be insufficiently sensitive to detect true curvature. What is seen may be due to random error.

It may be appropriate to calculate a pooled error term comprising the residual error plus all deviations from linear regression. This possibility will be explored in the "Alternative Calculation" section later in this chapter.

The value of variance ratio for preparations is within the warning limit arbitrarily set by some authorities. In previous examples, which were all symmetrical assays, this criterion was dismissed as being of little importance. In the case of this and other asymmetrical assays, a high variance ratio for preparations might be cause for concern as bias (always negative) increases as potency ratio estimates depart from unity. More important than the size of the variance ratio is the extent of deviation of the potency ratio from unity. As a crude guide, when estimated potency ratio is between 0.8 and 1.25, bias in the range 0.0 to –1 percent may be expected. (This topic is discussed in Chapter 11).

The assessment of the analysis of variance is satisfactory, and so confidence limits for the potency estimate may be calculated. Calculate the term S_{xx}, the sum of the squares of deviations of individual log doses from the mean log dose. The dose ratio is 5:4, so the log dose interval is 0.096910. The deviations from the mean log dose are one log-dose interval (dose levels 2 and 4) and two log-dose intervals (dose levels 1 and 5); the replication is 9 so that S_{xx} is obtained as:

$$9\left[(2 \times 0.0969910)^2 + (0.0969910)^2 + (-0.0969910)^2 + (-2 \times 0.0969910)^2\right]$$

$$= 0.8452$$

Calculate g, the index of significance of the slope b, using Equation (9.22) as

$$g = \frac{747.8 \times 2.021^2}{816.91^2 \times 0.8452} = 0.005415$$

where s^2, b^2, and S_{xx} have the values already calculated, and t is Student's t obtained from tables for $P = 0.95$ and 40 degrees of freedom as 2.021.

As g is less than 0.1, calculate confidence limits of the potency estimates by the so-called approximate method.

Calculate $V(M_1)$, the variance of the logarithm of the potency estimate of the first unknown, by substituting the known values in Equation (9.23)

$$V(M_1) = \frac{747.8}{816.91^2}\left[\frac{1}{36} + \frac{1}{9} + \frac{-0.024483^2}{0.8452}\right] =$$

$$0.00112056[0.027778 + 0.111111 + 0.000709] = 0.000156428$$

The standard error of the logarithm of the potency estimate, s_m, is then obtained by Equation (9.24) as:

$$s_m = (0.000156428)^{0.5} = 0.0125071$$

Log percent confidence limits are given by Equation (9.25) as

$$2 \pm 2.021 \times 0.0125071 = 1.9747 \text{ and } 2.0253$$

The corresponding percentage limits are 94.3 and 106.0% ($P = 0.95$).

Report

To express findings in terms of conformity with labeled claim, that is, the weight of kanamycin base in a vial of average weight, assume that 1000 IU of kanamycin activity corresponds to 1 mg of kanamycin base. (The *British Pharmacopoeia, BP* does not include any statement as to the potency of kanamycin base.)

The unknown test solution had a concentration of 6.1525 µg/ml (0.0061525 mg/ml); its potency relative to that of the standard reference solution was found to be 0.9452. The factor relating actual standard reference solution potency to its nominal potency was 0.9991. Therefore, potency of the solid material was estimated to be

$$\frac{5.000 \times 0.9991 \times 0.9452}{0.0061525} = 767.45 \text{ IU/mg}$$

and the kanamycin activity in a vial of average weight 1284 mg was

$$767.45 \times 1284 = 985,406 \text{ IU}$$

equivalent to 985.4 mg of kanamycin base.

Applying the Confidence Limits

A vial of average content 1284 mg contains 985.4 mg of kanamycin base, limits 929.2 to 1044.5 mg ($P = 0.95$).

ALTERNATIVE CALCULATION

In view of the suggestion that the components for different degrees of curvature may be due to random error, it is perhaps more appropriate to combine all deviations from linear regression with residual error to form a *pooled error* term, thus:

$$\text{pooled error} = 1213.3 + 263.5 + 239.0 + 29,911.1 = 31,626.9$$

Pooled error has 43 degrees of freedom, leading to a mean squares value of

$$31,626.9/43 = 735.5$$

Applying the usual calculations [Equation (9.23) and Equation (9.24)]

s_m becomes 0.01230667

t becomes 2.016 $\left(\text{for } P = 0.95 \text{ and } 43 \text{ degrees of freedom}\right)$

leading to log percent confidence limits of

$$2 \pm 0.01230667 \times 2.016 = 1.975190 \text{ to } 2.024810$$

corresponding to percent confidence limits of 94.4 to 105.9%. This alternative calculation has given results differing only very little from the original calculation.

REFERENCES

Brownlee, K.A., Delves, C.S., Dorman, M., Green, C.A., Grenfell, E., Johnson, J.D.A., and Smith, N., 1948, *J. Gen. Microbiol.*, 2, 40.

Brownlee, K.A., Loraine, P.K., and Stephens, J., 1949, *J. Gen. Microbiol.*, 3, 347.

Finney, D.J. 1978. *Statistical Method in Biological Assay*. London: Griffin.

Humphrey, J.H., Lightbown, J.W., Mussett, M.V., and Perry, W.L.M. 1953. The international standard for aureomycin, *Bull. World Hlth. Org.*, 9, 851.

Lees, K.A. and Tootill, J.P.R. 1955a. Microbiological assay on large plates. Part I General considerations with particular reference to routine assays, *Analyst*, 80, 95.

Lees, K.A. and Tootill, J.P.R. 1955b. Microbiological assay on large plates. Part II, Precise assay, *Analyst*, 80, 110.

Lees, K.A. and Tootill, J.P.R. 1955c. Microbiological assay on large plates. Part IV, High throughput, low precision assays, *Analyst*, 80, 531.

The British Pharmacopoeia, 1993.

The European Pharmacopoeia, Second Edition, 1993.

The European Pharmacopoeia, Third Edition, 1997.

The European Pharmacopoeia, Third Edition, Supplement 2000.

The European Pharmacopoeia, Third Edition, Supplement 2001.

The International Pharmacopoeia, Third Edition, 1979.

The International Pharmacopoeia, Second Edition, 1967.

The United States Code of Federal Regulations.

The United States Pharmacopoeia, 24 in [111].

10 Slope Ratio Assays — Some Designs and Their Evaluation

INTRODUCTION

It is generally in the case of turbidimetric assays of vitamins and amino acids that the slope ratio assay is applicable. There are official assay procedures described in the *USP, United States Pharmacopoeia*, but not in the *EP, European Pharmacopoeia*. Despite their absence from monographs, the *EP, European Pharmacopoeia* does give guidance on the evaluation of slope ratio assays.

SOME POSSIBLE DESIGNS

The ideal concepts on which slope ratio assays are based are:

1. Zero response to zero dose
2. Responses that are directly proportional to dose

Thus, the simplest possible design would consist of:

1. A zero dose or blank test solution containing no active ingredient
2. A test solution consisting of a single known concentration of a reference standard
3. A test solution prepared from a single defined concentration of the unknown preparation to be assayed

Such a test could be described as a *three-point common-zero assay* and would be the most efficient possible assay. It suffers from the drawback that no checks on validity would be possible. That is, it would not be possible to demonstrate (1) that the dose-response lines were rectilinear and (2) that they coincided at zero dose. In fact, this design does not appear to be used at all in microbiological assay, although it could be useful for routine purposes if prior knowledge of the assay showed that it did lead to a valid potency estimate.

By way of contrast, this design is very common in analytical chemistry, although the term *three-point common-zero assay* is not applied. An ultra violet spectrophotometric assay based on published information of the absorbance characteristics of the substance assayed could be described as a three-point common-zero assay.

The addition of another dose level of both reference standard and unknown leads to the five-point common-zero assay. This design is well established in microbiological assay and can demonstrate whether there is coincidence of the dose-response lines at zero dose.

There are many other possibilities. Of symmetrical assays, there are the three-dose level or seven-point common-zero assay, and the four-dose level or nine-point common-zero assay, both of which can demonstrate rectilinearity of the dose-response line or deviations from that ideal.

The *European Pharmacopoeia* (1997) indicates a preference for the five-point common-zero assay. This is a compromise between greatest efficiency and minimal tests for conformity with assay principles. It is a very good compromise.

However, the 2000 supplement to the *EP* makes reference to designs with both three and four dose levels of both preparations and omitting the blank, or zero, dose controls from the calculations. The examples given are not of *micro*biological assays. Although this does not detract from the value of these examples in *illustrating* assay evaluation, the fact remains that the two-dose level assay is a more efficient design. This is true whether or not the zero-dose responses are used in the calculation.

Asymmetrical assays typically consist of several dose levels of reference standard (perhaps five or six levels) and one or two levels of the unknown preparation to be assayed. These are useful in the assay of samples for which the analyst has little idea of the potency of the sample before testing.

The standard response line would generally include a wide range of doses and would probably be curved. In such cases, sample potency could be estimated by graphical interpolation from the standard response line. Arithmetical procedures leading to estimation of confidence limits are inappropriate when the response line is curved. When a precise potency estimate of such samples is required, the estimate obtained by interpolation may be used as the starting point for a further assay using a symmetrical design.

PRELIMINARY EVALUATION

The initial inspection of raw data should follow the general guidance given in Chapter 8. It is useful to draw a graph. It is a fact that sometimes the observed response to zero dose is not collinear with the responses to actual doses of reference standard and unknown but the two response lines extrapolated to zero dose do coincide on the *y* axis. In such cases the actual response to zero dose may be omitted from the calculation. It is important to remember that (in contrast to parallel-line assays) curvature of the response lines may seriously bias the potency estimate by the arithmetical procedure unless sample and standard potencies are quite close. It is not easy to quantify the extent of bias arising from curvature; previous experience of the assay may help. Although the final statistical evaluation will provide a figure for significance of curvature, this will not be of indisputable value in assessing the true worth of the assay. The figure for significance of curvature may suggest that the assay is invalid, but if sample and standard responses were close enough to suggest a potency estimate in the range 95 to 105%, this may well be a good estimate. This is an area where common sense and the experience of the analyst should prevail. Unfortunately, it may be hard to convince regulatory authorities that this is so.

Finally, if the sample and standard responses are not close and response lines are curved, then the assay may be salvaged by interpolation from the standard curve.

GENERAL STATISTICAL PROCEDURES

Evaluation procedures are generally based on multiple linear regression. When there is just one unknown, the equation (for both lines) is

$$y = a + b_s v_s + b_u v_u \qquad (10.1)$$

where y is the response, v_s and v_u are doses of standard and unknown, respectively, and b_s and b_u are the slopes of standard and unknown response lines, respectively.

The slopes for standard and unknown are not independent because responses to the zero-dose control are taken into consideration in calculating them. The multiple linear regression calculation of slopes ensures that both have a common origin (a of Equation 10.1), which is the best estimate of the response to zero dose based on all responses. This does not necessarily coincide with the observed mean of the responses to zero dose alone.

The *European Pharmacopoeia* (1997) describes an evaluation procedure for the five-point common-zero assay and refers the reader to Clarke (1952) and Barraclough (1955) for the assay of two or more preparations. The calculation procedures of the 2000 supplement to the *EP* differ from those of earlier editions but appear to offer no advantages.

An evaluation procedure was described by Bliss (1951) based on multiple linear regression. In contrast to the *European Pharmacopoeia* procedure, that of Bliss can be adapted readily to assays having more than two dose levels including asymmetric assays. Both the Bliss and *European Pharmacopoeia* procedures are complex. The Bliss procedure minimizes the use of rule-of-thumb expressions and so is preferred by this writer. The formulas for calculating degrees of freedom given in the *EP* are perhaps designed to minimize errors being made. The downside is that they absolve the analyst from the need to indulge in a very simple thought process and thus mitigate understanding of the evaluation process.

Only the Bliss procedure will be illustrated in detail here. However, the results of the analysis of variance by the two procedures will demonstrate that they are mutually compatible.

THE BLISS PROCEDURE

The calculation is lengthy but may be made more readily understandable by breaking it down into its component steps. First, the nomenclature is defined (the five-point common-zero assay is used here to illustrate nomenclature).

x_s = the value of an individual dose of standard; the doses of standard (and unknown) are coded; thus in the five-point common-zero assay, dose level 1 is coded as 0.5 and dose level 2 as 1.0

x_u = similarly, the (coded) value of an individual dose of unknown

y = a value of an individual response

r = the replication

N = the total number of responses to standard, unknown, and zero doses

Sy = the grand total of all observed responses

The calculation:

Step 1. Calculate two basic functions:

Sx_s is the sum of the coded log doses for the standard and so is calculated as

$$Sx_s = r(0.5 + 1.0) \qquad (10.2)$$

Sx_u is the sum of the coded log doses for the unknown and is calculated as

$$Sx_u = r(0.5 + 1.0) \qquad (10.3)$$

Note: Coding could alternatively be 1 and 2. Here, 0.5 and 1.0 are used so as to be consistent with the *European Pharmacopoeia.*

Step 2. Calculate the values of three functions of dose:

$$1. \quad Sx_s x_s = S(x_s)^2 - (Sx_s)^2 / N \qquad (10.4)$$

$$2. \quad Sx_u x_u = S(x_u)^2 - (Sx_u)^2 / N \qquad (10.5)$$

$$3. \quad Sx_s x_u = \left[S(x_s)(x_u) \right] - \left[(Sx_s)(Sx_u)/N \right] \qquad (10.6)$$

Step 3. Calculate the values of two functions of dose and response:

$$4. \quad Sx_s y = \left[S(x_s y) \right] - \left[(Sx_s)(Sy)/N \right] \qquad (10.7)$$

$$5. \quad Sx_u y = \left[S(x_u y) \right] - \left[(Sx_u)(Sy)/N \right] \qquad (10.8)$$

The meanings of these two expressions will be made clear in the example that follows in the section below, "Evaluation of a Two-Dose Level Assay Using the Bliss Procedure."

Step 4. Simplifying the nomenclature, put

$$Sx_s x_s = p$$

$$Sx_u x_u = q$$

$$Sx_s x_u = r$$

$$Sx_s y = s$$

$$Sx_u y = t$$

Note: distinguish between r here and r given earlier in this chapter for replication.

Step 5. Now substitute these values in the expressions for b_s and b_u, the slopes of the response lines for standard and unknown, respectively. These two expressions are the solutions to two simultaneous equations:

$$b_s = (sq - tr)/(pq - r^2) \tag{10.9}$$

$$b_u = (tp - rs)/(pq - r^2) \tag{10.10}$$

Step 6. Obtain the potency ratio R as

$$b_u/b_s \tag{10.11}$$

Statistical Evaluation

Step 1. Carry out an analysis of variance to determine the residual error, s^2.

The procedure is the same as that for parallel line assays except for the calculation of regression squares, which is by the equation:

$$\text{regression squares} = [b_s \times Sx_s y] + [b_u \times Sx_u y] \tag{10.12}$$

Squares for deviations from regression are obtained as the difference:

preparation squares – regression squares

Squares for residual error are obtained as the difference:

total squares – preparation squares

Step 2. Calculate g, the index of significance of the slope, using the expression:

$$g = \frac{s^2 \times t^2 \times Sx_u x_u}{b_s^2 \left[\left(Sx_s x_s \right)\left(Sx_u x_u \right) - \left(Sx_s x_u \right)^2 \right]}$$

(10.13)

Step 3. When g is less than 0.1 (as it invariably is in any assay that is of any value), calculate the variance of the potency ratio, $V(R)$, using the simplified expression:

$$V(R) = \left[\frac{s^2}{b_s^2} \right] \times \left[\frac{Sx_s x_s + 2RSx_s x_u + R^2 Sx_u x_u}{\left(Sx_s x_s \right)\left(Sx_u x_u \right) - \left(Sx_s x_u \right)^2} \right]$$

(10.14)

Step 4. Obtain the standard error of the potency ratio, s_R as

$$s_R = [V(R)]^{0.5}$$

(10.15)

Step 5. Proceed to the calculation of confidence limits as

$$R \pm t \times s_R$$

(10.16)

EVALUATION OF A TWO-DOSE LEVEL ASSAY USING THE BLISS PROCEDURE

A niacin assay carried out in 1994 will be used to illustrate this calculation. The raw data are presented in Table 10.1.

EXAMPLE 10.1

The assay of niacin using a five-point common-zero assay design.

Test organism: *Lactobacillus arabinosus* NCIB 8030

Standard dose levels: 0.01 and 0.02 µg/ml

Sample dose levels, nominally: 0.01 and 0.02 µg/ml

As mentioned in the section on preliminary evaluation, often in slope ratio assays, responses to zero dose are not collinear with responses to doses of standard and of unknown. Comparison of the slopes in Table 10.1 shows that such is the case in this example. The extent of deviation from collinearity is illustrated visually in

TABLE 10.1
The Raw Data from Example 10.1, a Turbidimetric Assay of Niacin Using a Five-Point Common-Zero Design

		Standard		Unknown	
Doses →	Zero	Low	High	Low	High
	21	364	720	332	624
	41	365	678	339	645
	15	391	697	368	629
	35	378	677	347	629
Treatment totals →	112	1498	2772	1386	2527
Treatment means →	28.0	374.5	693.0	346.5	631.8

Slopes based on mean responses:

standard low – zero	=	346.5
standard high – standard low	=	318.5
unknown low – zero	=	318.5
unknown high – unknown low	=	285.3

Note: Recorded observations are 1000 × optical absorbance at 580 nm.

Figure 10.1 and Figure 10.2. In Figure 10.1, lines of best fit are plotted based on all mean responses; in Figure 10.2, lines of best fit are plotted excluding mean response to zero dose. It appeared from the graphs that deviation was not serious. In cases where the mean zero dose response deviates significantly from collinearity, it is permissible to omit it from the calculation.

It was shown in the statistical evaluation that follows that the deviation was not significant and so retention of zero-dose responses was justified.

Step 1. Using coded doses 0.5 and 1.0 for low- and high-dose levels, respectively, calculate the basic functions of dose by using Equation 10.2 and Equation 10.3 as:

$$Sx_s = 4(0.5 + 1.0) = 6$$

$$Sx_u = 4(0.5 + 1.0) = 6$$

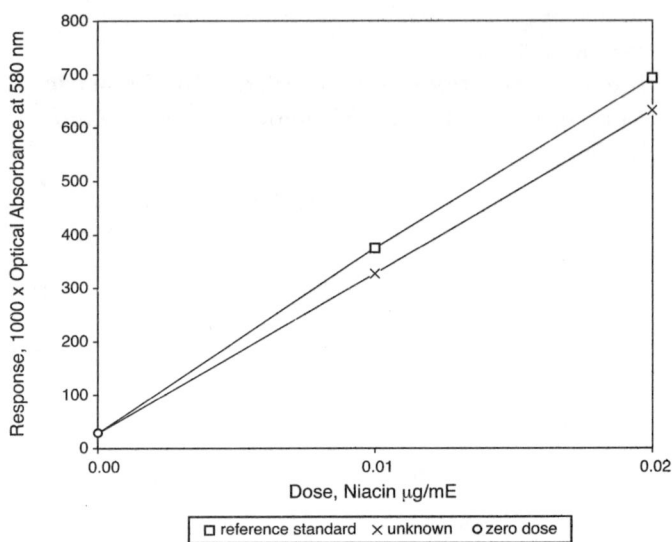

FIGURE 10.1 A graphical representation of all mean responses in Example 10.1. Responses to zero dose deviate slightly from being collinear with those to dose levels 1 and 2. The slight curvature of the response lines is just detectable on careful inspection. However, compare with Figure 10.2.

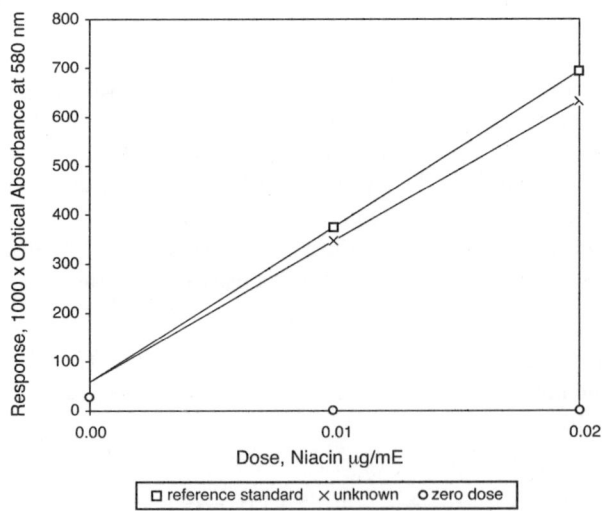

FIGURE 10.2 A graphical representation of the mean responses to dose levels 1 and 2 of reference standard and unknown in Example 10.1. The two response lines coincide with the y axis (zero dose) at a value of 59 whereas the observed mean response to the zero-dose controls was 28. This does not appear to be a serious discrepancy but serves to illustrate the principles on which it might be appropriate to omit from the calculations the actual responses to zero dose. ·

Step 2. Calculate the values of the three functions of dose using Equation 10.4 to Equation 10.6.

1. $Sx_s x_s = p = 4(0.5^2 + 1.0^2) - 6^2/20 = 3.2$

2. $Sx_u x_u = q = 4(0.5^2 + 1.0^2) - 6^2/20 = 3.2$

3. $Sx_s x_u = r = 0 - (6 \times 6)/20 \qquad = -1.8$

Note 1. Whenever x_s has a value other than zero, x_u is zero, and vice versa; thus the value of the first term of expression 3 is always zero.

Note 2. For this assay design, with this coding of doses and replication of four, the values of expressions 1, 2, and 3 will always be as just noted.

Step 3. Now using Equation 10.7 and Equation 10.8 and the raw data of Table 10.1, calculate two further functions related to dose and response:

4. $Sx_s y = s = [(0.5 \times 1498) + (1.0 \times 2772)] - [(6 \times 8295)/20]$

$$= 3521.0 - 2488.5 = 1032.5$$

5. $Sx_u y = t = [(0.5 \times 1386) + (1.0 \times 2527)] - [(6 \times 8295)/20]$

$$= 3220.0 - 2488.5 = 731.5$$

Steps 4 and 5. Now changing to the simplified nomenclature, substitute these values in Equation 10.9 and Equation 10.10 to obtain the slopes for standard and unknown:

$$b_s = \frac{(1032.5 \times 3.2) - (731.5 \times -1.8)}{(3.2 \times 3.2) - (1.8)^2} = 660.1$$

$$b_u = \frac{(731.5 \times 3.2) - (1032.5 \times -1.8)}{(3.2 \times 3.2) - (1.8)^2} = 599.9$$

Step 6. Obtain the potency ratio R by Equation 10.11 as:

$$b_u/b_s = 599.9/660.1 = 0.9088$$

Statistical Evaluation

The values for total deviation squares and treatment squares are calculated in exactly the same way as for parallel-line assays, thus:

Step 1. Total deviation squares (not written in full)

$$\left[21^2 + 364^2 + \ldots + 347^2 + 629^2\right] - \left[8295^2/20\right]$$

$$= 4,710,984 - 3,479,448.2 = 1,124,590.0$$

Step 2. Treatment sums of squares

$$\left[\left(112^2 + 1498^2 + 2722^2 + 1386^2 + 2527^2\right)/4\right] - \left[8295^2/20\right]$$

$$= 4,561,814 - 3,440,351 = 1,121,463$$

Step 3. Regression sums of squares is given by Equation 10.12 as

$$[660.1 \times 1032.5] + [599.9 \times 731.5] = 1,120,380.1$$

Step 4. Total deviations from regression are obtained as the difference:

treatment squares – regression squares, thus:

$$1,121,463 - 1,120,380.1 = 1082.9$$

The analysis of variance is summarised in Table 10.2, in which residual (error) variation is obtained by difference.

TABLE 10.2
Summary of Analysis of Variance (1) for Example 10.1, a Five-Point Common-Zero Slope Ratio Assay

Source of Variance	d.f.	Sum of Squares	Mean Squares	Variance Ratio	Limiting Value	At Arbitrary Probability Level, *P*	Calculated Probability
Regression	2	1,120,380	560,190.0	2687.19	>10.16	0.001	<0.0001
Deviations from regression	2	1083	541.5	2.60	<3.52	0.050	0.1073
Treatments	4	1,121,463					
Residual error	15	3127	208.5	1.00			
Total	19	1,124,590					

Note: Limiting values in column 6 refer to the corresponding arbitrary probability level of column 7.

Degrees of freedom are obtained thus:

> There are 5 treatments giving $5 - 1 = 4$ d.f.
> There is 1 d.f. for each regression line, that is 2 d.f.

Degrees of freedom for deviations from regression are obtained as:

> d.f. for treatments − d.f. for regression $= 4 - 2 = 2$

Total degrees of freedom are number of observed responses −1, that is $20 - 1 = 19$
Degrees of freedom for residual error are obtained as:

> Total d.f. − treatments d.f. $= 19 - 4 = 15$.

Mean squares and variance ratios are calculated for regression and deviations from regression.

Discussion of Assay Validity Based on the Analysis of Variance

The variance ratio for deviations from regression (2.60) is less than the limiting value, 3.52, so these deviations are not significant and there seems to be no cause to reject the responses to zero dose. The calculation is continued, leading to confidence limits of the potency estimate.

Step 5. Now, using Equation (10.13), calculate g, the index of significance of the slope as:

$$g = \frac{208.5 \times 2.145^2 \times 3.2}{660.1^2 \times \left[(3.2 \times 3.2) - (-1.8)^2\right]} = 0.000993$$

Step 6. As g is less than 0.1, calculate variance of R by the simplified formula Equation (10.14) as

$$V(R) = \left[\frac{208.5}{660.1^2}\right] \times \left[\frac{3.2 + (2 \times 0.9088 \times -1.8) + \left(0.9088^2 \times 3.2\right)}{(3.2 \times 3.2) - (-1.8)^2}\right]$$

$$= 0.0001758$$

Step 7. Obtain standard error of the potency estimate, s_R, by Equation (10.15) as

$$0.0001758^{0.5} = 0.01326$$

Step 8. Finally, obtain confidence limits by Equation (10.16) as

$$0.9088 \pm (0.01326 \times 2.145)$$

or

$$0.8806 \text{ to } 0.9371$$

report as

$$0.881 \text{ to } 0.937 \ (P = 0.95)$$

that is, ± 3.1 percent of potency estimate.

THE PROCEDURE OF THE *EUROPEAN PHARMACOPOEIA*

The calculation procedure of the *European Pharmacopoeia*, Second Edition, 1993 remained unchanged in the 1997 version. However, there were radical changes in the Third Edition, Supplement 2000. The latter procedure was applied to the data of Example 10.1. Rather than subject the reader to the tedium of the detailed calculation, only the results of the analysis of variance are shown here. These are presented in Table 10.3. Estimated potency and confidence limits were identical with those obtained from Table 10.2.

TABLE 10.3
Summary of Analysis of Variance (2) for Example 10.1, a Five-Point Common-Zero Slope Ratio Assay

Source of Variance	d.f.	Sum of Squares	Mean Squares	Variance Ratio	Limiting Value	At Arbitrary Probability Level, *P*	Calculated Probability
Regression	2	1,120,380	560,190.0	2687.19	>10.16	0.001	<0.0001
Blanks	1	1072	1072.0	5.14	<4.38	0.050	0.0386
Intersection	1	11	11.0	0.05	<4.38	0.050	0.8261
Subtotal	4	1,121,463					
Treatments	4	1,121,463					
Residual error	15	3127	208.5	1.00			
Total	19	1,124,590					

Note: Limiting values in column 6 refer to the corresponding arbitrary probability level of column 7.

DISCUSSION OF ASSAY VALIDITY BASED ON THE ANALYSIS OF VARIANCE

The summary of analysis of variance presented in Table 10.3 differs from that of Table 10.2 only in the fact that instead of a pooled value for deviation from regression, the individual components are shown as sum of squares for blanks and sum of squares for intersection. These are 1072 and 11 respectively. Their sum, 1083, is identical with the pooled value shown in Table 10.2. These individual values give variance ratios of 5.14 and 0.05 respectively. The former is a little above the arbitrary limiting value of 4.38 for $P = 0.05$ and so suggests that it might have been better to omit responses to zero dose.

THE BLISS PROCEDURE APPLIED OMITTING RESPONSES TO ZERO DOSE

As the calculation of the section "Evaluation of a Two-Dose Level Assay Using the Bliss Procedure" suggested that it might have been better to omit responses to zero dose, this was done but using the preferred Bliss calculation. Details of the calculation are not given here. The results of the analysis of variance are presented in Table 10.4, from which it is seen that now deviations from regression are negligible. The estimated potency ratio from this calculation was 0.905 and confidence limits ($P = 0.95$) were 96.5 to 103.5%.

AN OVERVIEW

The omission of the responses to zero dose has led to a statistical analysis in which there are no question marks as to validity of the assay. The estimated potency ratio is 0.9050 instead of 0.9088 — a difference of a mere 0.42%. Confidence limits have

TABLE 10.4
Summary of Analysis of Variance (3) for Example 10.1, a Five-Point Common Zero Slope Ratio Assay

Source of Variance	d.f.	Sum of Squares	Mean Squares	Variance Ratio	Limiting Value	At Arbitrary Probability Level, P	Calculated Probability
Regression	2	373,574.2	186,787.1	833.03	>12.97	0.001	<0.0001
Deviation from regression	1	11.0	11.0	0.05	<4.75	0.050	0.8268
Treatments	3	373,585.2					
Residual error	12	2690.7	224.2	1.00			
Total	15	376,275.9					

Note: Limiting values in column 6 refer to the corresponding arbitrary probability level of column 7.

widened from ± 3.1 to ± 3.5% — again not a big difference. It is suggested here that as a routine practice it would be preferable to omit the potentially troublesome responses to zero dose and thus obtain a result that is less likely to be questioned as to its statistical validity. Perhaps it would be more economical to omit zero dose controls from the practical design.

A final thought: whilst methods of presentation change, the fundamental mathematical truths established by statisticians half a century ago remain unchanged.

REFERENCES

Barraclough, C.G. 1955. *Biometrics*, 11, 286.

Bliss, C.J. 1951. In *The Vitamins*, Vol II, edited by P. Gyorgy. New York: Academic Press.

Clarke, P.M. 1952. *Biometrics*, 8, 370.

The European Pharmacopoeia (EP), 1993. Second Edition, Strasbourg: Council of Europe.

The European Pharmacopoeia (EP), 1997. Third Edition. Strasbourg: Council of Europe.

The European Pharmacopoeia (EP), 2000. Supplement to the Third Edition. Strasbourg: Council of Europe.

The United States Pharmacopoeia (USP), National Formulary 24, <111>, Rockville, MD: U.S. Pharmacopoeial Convention.

11 Choice of Experimental Design

INTRODUCTION

In Chapters 9 and 10, some designs of assay that are commonly used worldwide were seen. In this chapter, a wider range of designs will be introduced, and their suitability for use in different circumstances will be considered. The main emphasis is on parallel-line assays; similar principles generally apply to slope ratio assays.

Assays of pharmaceuticals are generally subject to scrutiny by regulatory bodies, and so analysts must take heed of regulatory guidance. It is important to note that, despite regulatory guidance, analysts do have freedom to select experimental designs of their own choice, provided that they can demonstrate that these are as good as, or better than, those described in official compendia.

As a first step, we need to consider what designs are available. This is discussed in the next section for agar diffusion assays, and in later sections for turbidimetric assays of growth-inhibiting substances and for turbidimetric assays of growth-promoting substances.

AVAILABLE EXPERIMENTAL DESIGNS FOR AGAR DIFFUSION ASSAYS

The possible experimental designs for plate assays are determined by the size and shape of the plate. Commonly available plates are petri dishes, which can accommodate six zones, and large square plates, which can accommodate 36 or 64 zones. Less common are larger plates for 81 and 144 zones. The possibilities with the aforementioned plates are shown in Table 11.1.

NOMENCLATURE

Although much of the nomenclature used in Table 11.1 and elsewhere refers specifically to parallel-line assays, some terms are equally applicable to slope ratio assays. It will be self-evident which are of general application.

A parallel-line assay is an assay in which the assumption on which the calculation is based is that when response to standard is plotted against logarithm of dose, and response to unknown is plotted against logarithm of its nominal dose, the result will be two straight parallel lines. Naturally, when only two-dose levels are used, there is no internal assay evidence that

TABLE 11.1
Some Designs for Agar Diffusion Assays

Plate Dimensions	Design	Number of Dose Levels	Number of Preparations	Degree of Replication
10 cm diam	2 + 2	2	3	n[a]
10 cm diam	3 + 3	3	2	n[a]
10 cm diam	5 + 1[b]	5	m[c]	n[a]
24 × 24 cm	2 + 2, 6 × 6 LS[d]	2	3	6
24 × 24 cm	3 + 3, 6 × 6 LS	3	2	6
30 × 30 cm	2 + 2, 8 × 8 LS	2	4	8
30 × 30 cm	2 + 2, 8 × 8 LS	2	2	16
30 × 30 cm	2 + 2, 8 × 8 QLS[d]	2	8	4
30 × 30 cm	8 × 8 random[e]	4	13	4
34 × 34 cm	3 + 3, 9 × 9 LS	3	3	9
44 × 44 cm	3 + 3, 12 × 12 LS	3	4	12

[a]   The degree of replication n can be increased to any number by using more plates.

[b]   These are asymmetrical designs.

[c]   The number of preparations m can be increased from 2 to any number by adding one more set of plates for each additional unknown.

[d]   The abbreviations LS and QLS represent Latin square and Quasi-Latin Square, respectively.

the lines are straight. Similarly, when only one dose level of unknown is used, there is no internal assay evidence that the lines are parallel. Nevertheless, previous knowledge is that the lines would be parallel and the calculation assumes two straight and parallel lines. Thus, an assay having five dose levels of standard and only one dose level of each unknown is described here as a parallel-line assay. Normally, dose levels form a geometrical progression so that their logarithms form an arithmetical progression. (Note that although the *United States Pharmacopoeia* does describe calculation procedures for dose levels not forming a geometrical progression, these appear to be little used.)

A preparation is a standard or an unknown. In an assay in which three unknowns are compared with one standard, there are four preparations.

Treatment corresponds to *test solution*. In an assay having one standard and two unknowns, each at three-dose levels, there are nine treatments; in an assay having one standard and one unknown, each at two-dose levels but with duplicate weighings of each preparation, there are eight treatments.

The notation 2 + 2 indicates two-dose levels of each of standard and unknown. This description applies both to assays of a single unknown and to multiple assays.

The notation 3 + 3 indicates three-dose levels of each of standard and unknown. This description applies both to assays of a single unknown and to multiple assays.

(Note that elsewhere a different meaning has been attached to this form of expression, for example, 1 + 3 has been used to indicate an assay in which three unknowns are compared with one standard and not to indicate one and three-dose levels, respectively, for two preparations.)

The notation 6×6 (also known as 6 by 6) indicates the six rows and six columns of a Latin square design; 8×8, 9×9, and 12×12 have analogous meanings. These two forms of description are not mutually exclusive; thus, a 6×6 assay will also be either 2 + 2 (for two unknowns) or 3 + 3 (for one unknown).

The notation LS means a randomized Latin square; in an $n \times n$ Latin square, each of the n numbers appears once (and only once) in each of the n rows and n columns.

The notation QLS means a randomized quasi-Latin square; in an $n \times n$ quasi-Latin square, positions are numbered 1 to 16. A random selection of eight of these numbers appears once (and once only) in four of the eight rows; the remaining eight numbers appear in the other four rows. Similarly, a random selection of eight of these numbers appears once (and once only) in four of the eight columns; the remaining eight numbers appear in the other four columns..

The notation R refers to a random arrangement of numbers. In the single example quoted in Table 11.1, test solutions are numbered 1 to 16 and distributed in a completely random manner; thus, two adjacent positions may have the same number. In this particular design, there are four-dose levels of standard and a single dose level of each of 12 unknowns, making 13 preparations and 16 treatments. Each treatment has a replication of four.

The term *symmetrical* describes an assay in which each preparation has the same number of dose levels and the ratio between dose levels is the same for each preparation; replication for each treatment is the same.

AN OUTLINE OF SOME DESIGNS NOT ILLUSTRATED BY EXAMPLES IN CHAPTER 9

1. *A two-dose level assay using petri dishes.* The standard size of petri dish can accommodate six inhibition zones, so when a two-dose level assay suffices, it is economical to compare two unknowns with the standard at the same time. Typically, six dishes would be used. In the analysis of variance, there would be a total 35 degrees of freedom, of which 5 would be allocated to treatments and 5 to plates, leaving 25 for residual error. The breakdown of the treatment degrees of freedom would be:

Regression	1
Preparations	2
Parallelism	2

   This design is appropriate when moderate precision is acceptable.

2. *A two-dose level assay using a square plate and 6 × 6 Latin square design.* This design is for the simultaneous assay of two unknowns. In the analysis of variance there would be a total 35 degrees of freedom, of which 5 would be allocated to treatments, 5 to rows, and 5 to columns, leaving 20 for residual error. The breakdown of the treatment degrees of freedom would be as in 1. Also like 1, this design is appropriate when moderate precision is acceptable.

3. *A two-dose level assay using a square plate and 8 × 8 Latin square design.* This design is for the assay of one unknown with a high replication. There are four treatments; high and low doses of reference standard and of unknown. Each treatment is allocated two numbers: 1 and 5, 2 and 6, 3 and 7, 4 and 8. Thus, each treatment appears twice in each row and each column. In the analysis of variance, there would be a total 63 degrees of freedom, of which 3 would be allocated to treatments, 7 to rows, and 7 to columns, leaving 46 for residual error. The breakdown of the treatment degrees of freedom would be

Regression	1
Preparations	1
Parallelism	1

The design is suitable when good precision is needed.

4. *A two-dose level assay using a square plate and 8 × 8 quasi-Latin square design.* This design permits the simultaneous assay of seven unknowns. Test solutions representing the 16 treatments are numbered 1 to 16 and applied to the plate in accordance with the quasi-Latin square design. Each test solution appears in half the rows and half the columns. There are only eight responses for each preparation, so precision is moderate to low. The analysis of variance is more complex than that of other examples that have been shown in this text. The reader is referred to the original publication by Lees and Tootill (1955a) or to the illustration of this design by Hewitt (1977a, 177–182).

5. *A large plate assay using a four-dose level standard curve.* This design, due to Lees and Tootill (1955b), is useful for screening large numbers of unknowns to obtain a rough estimate of potency. This assay compares responses to single doses of 12 unknowns with a four-dose level standard curve in which the dose intervals are 2:1, thus covering an overall range of 8:1. Thus there are 16 treatments, each having a replication of only four. Test solutions are distributed on an 8 × 8 plate in accordance with a completely random design. Use of a random design may lead to replicates of the same test solution having adjacent positions on the plate. It is suggested here that use of a quasi-Latin square design may be perfectly satisfactory even though there is no intention to separate components of variation due to position on the plate.

The design leads to imprecise results; the authors (Lees and Tootill) suggest ±10%. However, it is very economical and serves its intended purpose well, provided that its limitations are recognized. The authors do not illustrate its statistical evaluation. One example of its statistical evaluation is given by Hewitt (1977a, 183–187).

REGULATIONS AND OPTIONS

Analysts tend to be conservative in their choice of design and use only one or up to three designs. In the U.S., the design that appears to be almost universally used is the 5 + 1 design of the *United States Pharmacopoeia*, employing five dose levels of standard (with a 5:4 dose ratio between adjacent dose levels) and a single dose level of any number of unknowns. According to Dabbah (1995) of the USP Convention, this is the referee method in the U.S., and analysts prefer to use it rather than risk dispute with the regulatory authority. However, the use of other procedures is explicitly permitted by the *USP*, as indicated by the statement:

> "Proof that an assayed potency meets its required confidence limits may be based also upon other recognized biometric methods that have a precision equivalent to that of the methods outlined herein." (*USP 24* 2000, <111>).

In the United States, the *United States Pharmacopoeia* has legal status and is totally independent of the Food and Drug Administration (FDA) (Dabbah 1994).

In Europe, either a 3 + 3 or a 2 + 2 design is preferred. On the question of whether to use a two-dose level or a three-dose level assay, the *British Pharmacopoeia* states:

1. "In order to be able to assess the validity of the assay, use at least three different doses of the Standard preparation and of the substance being examined ..."
   but this is then modified by the footnote:
2. "In routine examinations when rectilinearity of the system has been demonstrated over an adequate number of experiments using a three-point assay, a two-point assay may be sufficient, subject to agreement by the control authority. However, in all cases of dispute a three-point assay, as described above, must be applied." (BP 1993, App. XIVA, A167.

In the introduction to Section 5.3. 'Statistical Analysis," the 2001 Supplement to the *EP* states:

> This chapter provides guidance for the design of bioassays prescribed in the *European Pharmacopoeia* and for analysis of their results.

and

> The methods of calculation described in this annex (appendix) are not mandatory for the bioassays which themselves constitute a mandatory part of the *European Pharmacopoeia*. Alternative methods may be used, provided that they are not less reliable than those described here.

Thus, these major pharmacopoeias permit the analyst to use initiative in selecting an assay design appropriate to the circumstances.

It is a simple matter to compare experimental designs and to assess how design and replication will influence precision. It is prudent, therefore, to make such comparisons and to choose a design that is the most efficient and economical for a specific purpose.

It is also a simple matter to demonstrate whether a design is as good as, or superior to, those described in the pharmacopoeias. Mathematical principles only are involved; there is no practical experimental work to be done.

FACTORS INFLUENCING WIDTH OF CONFIDENCE LIMITS

The main design factor influencing the width of confidence limits is degree of replication. Other design factors include: number of dose levels, overall dose range, assay symmetry, and proximity of the potencies of the standard and unknown. The way these factors affect width of confidence limits of parallel-line assays is illustrated in the sections that follow by means of the now well-known Equation (9.23) for the variance of M:

$$V(M) = \frac{s^2}{b^2}\left[\frac{1}{N_s} + \frac{1}{N_T} + \frac{M^2}{S_{xx}}\right]$$

Rewrite as

$$V(M) = A \times B \tag{11.1}$$

where

$$A = s^2/b^2 \tag{11.2}$$

and

$$B = \left[\frac{1}{N_s} + \frac{1}{N_T} + \frac{M^2}{S_{xx}}\right] \tag{11.3}$$

It will be seen that A is mainly a function of the quality of the assay as determined by practical aspects. A (desirable) high value of b is dependent on the innate diffusion characteristics of the active substance in the medium, temperature, and time of diffusion and density of inoculum, as was discussed in Chapter 2.

A (desirable) low value of s^2 is dependent on sharpness of zone edges and the ability to measure them but is also dependent on the number of observed responses.

B is a function of (1) the design and replication of the assay and (2) the value of M, the logarithm of the potency estimate. The interaction of these various factors will be considered in the sections that follow.

THE EFFECT OF REPLICATION

The extent of replication has a profound effect on width of confidence limits. Clearly, increased replication reduces the width of confidence limits. However, its mechanism is somewhat more complex than may appear at first sight.

The random error of an assay, as represented by the parameter error squares, is a function of the innate characteristics of the assay — in particular, the clarity of the zone edge and the ability of the zone measuring system to detect it. It is independent of assay design. It seems reasonable to suppose that the size of error squares (although subject to random error) would be directly proportional to the number of observed responses.

Somewhat similarly, the slope b is dependent on the physical characteristics of the assay system only and so is independent of assay design. On the basis of these assumptions, we can explore the effect of assay design and replication on precision of the assay. Now we need to explore the influence of replication on the input and output values of these expressions.

Increase in replication has a twofold action in narrowing the confidence limits of the potency estimate:

1. It changes the number of degrees of freedom associated with residual error squares, and this leads to a reduction in the value of error mean squares.
2. Values of t decrease in accordance with the increased number of degrees of freedom for residual error.

It has been seen in examples in Chapter 9 that, in many assays, M is very small and so for our immediate purpose we can omit the term M^2/S_{xx} and write the approximate equation:

$$B = \left[\frac{1}{N_s} + \frac{1}{N_T} \right]$$

(11.4)

Starting with some arbitrary but realistic figures for error squares and slope, then substituting these values in Equation (11.1), Equation (11.2), and Equation (11.4) for different assay designs, we can arrive at a series of realistic ranges of confidence limits, which will reflect the practical situation.

A petri dish assay of one unknown compared with a standard at three-dose levels might, typically, consist of six plates, each having six zones (a replication of six). To explore the effect of different levels of replication, take this six-plate assay as the basis, input the realistic values for error squares and slope:

error squares is 0.8 so that with 25 degrees of freedom error mean squares is

$$0.8/25 = 0.032$$

slope $= 8.0$

then apply the appropriate modifications arising from changing replication. Changes arising include:

The value of error squares, and
The number of degrees of freedom associated with the residual error term of the analysis of variance, leading to:
The value of error mean squares, and
The value of t.

The values of these parameters arising for different numbers of plates are summarized in Table 11.2.

First taking six plates as an illustration (and as a reference point), we see from Table 11.2 that the value for error squares is 0.8; dividing by the corresponding number of degrees of freedom, 25, gives the value 0.032 for error mean squares. As the slope b is 8 mm regardless of number of plates, from Equation (11.2):

$$A = 0.032/8^2 = 0.0005$$

For six plates, $N_S = N_T = 18$, so that from Equation (11.4):

$$B = (1/18 + 1/18) = 0.111111$$

TABLE 11.2
Summary of Data Used to Assess the Influence of Replication on the Width of Confidence Limits

Number of Plates	d.f. for Residual Error	Error Squares	Error Mean Squares	Number of Zones per Preparation	t
2	5	0.267	0.0533	6	2.571
4	15	0.533	0.0356	12	2.131
6	**25**	**0.800**	**0.0320**	**18**	**2.060**
8	35	1.067	0.0305	24	2.030
10	45	1.333	0.0296	30	2.014
12	55	1.600	0.0291	36	2.004
14	65	1.867	0.0287	42	1.996
16	75	2.133	0.0284	48	1.992

Note: These data refer to a 3 + 3 assay having a 2:1 dose ratio.

and from Equation (11.1)

$$V(M) = 0.00050 \times 0.111111 = 0.000055556$$

and from Equation (9.24)

$$s_M = [0.000055556]^{0.5} = 0.007454$$

$t(P = 0.05)$ corresponding to 25 degrees of freedom $= 2.060$

Then by Equation (9.25), log percent confidence limits are

$$2 \pm 0.007454 \times 2.060 = 1.98465 \text{ and } 2.01535$$

giving confidence limits of 96.53 to 103.60%.

Now carrying out the analagous calculation for 16 plates, we see from Table 11.2 that the value for error squares is 2.133; dividing by the corresponding number of degrees of freedom, 75, gives the value 0.0284 for error mean squares. As the slope b is 8 mm regardless of number of plates, from Equation (11.2)

$$A = 0.0284/8^2 = 0.0004444$$

For sixteen plates, $N_S = N_T = 48$, so that from Equation (11.4)

$$B = [1/48 + 1/48] = 0.04167$$

and from Equation (11.1)

$$V(M) = 0.0004444 \times 0.04167 = 0.000018515$$

and from Equation (9.24)

$$s_M = [0.000018515]^{0.5} = 0.004303$$

$t(P = 0.05)$ corresponding to 75 degrees of freedom $= 1.992$

Then by Equation (9.25), log percent confidence limits are

$$2 + 0.004303 \times 1.992 = 1.99914 \text{ and } 2.0085$$

giving confidence limits of 98.05 to 101.99%.

The two ranges of confidence limits are

$$103.60\% - 96.53\% = 7.07\% \quad \text{(for six plates)}$$

$$101.99\% - 98.05\% = 3.94\% \quad \text{(for 16 plates)}$$

Thus, increasing replication from the typical six plates to 16 plates reduces the width of confidence limits by a factor of 3.94/7.07 = 0.557. Calculations such as these have lead to the figures presented in Table 11.3.

Substitution of other realistic values for error squares and for b would also lead to realistic but different values for widths of confidence limits. However, the values for their relative widths in Table 11.3 would be identical.

THE EFFECT OF THE VALUE OF M ON WIDTH OF CONFIDENCE LIMITS

It is an accepted principle of biological assays in general that the potency of the unknown test solution should be close to that of the reference standard. If they are not close, there are two possible consequences:

1. In the event of curvature of the response line, the responses to standard and unknown will be on different parts of the curve, leading to bias in the potency estimate.
2. The confidence limits of the potency estimate become wider as unknown and standard potency diverge.

The objective of this section is to quantify the latter.

Again, the three-dose level assay using six petri dishes will be used to illustrate the situation. For this purpose, we use Equation (11.1) and Equation (11.2) and explore the effect of changes in the value of M through the term M^2/S_{xx}. It is necessary to define the dose ratio as this determines the value of S_{xx}.

TABLE 11.3
The Influence of Replication on the Width of Confidence Limits (c.l.)

Number of Plates	$V(M)$	Confidence Limits $(P = 0.95)$	Relative Width of c.l.
2	0.000278	90.6 to 110.4%	2.78
4	0.000093	95.4 to 104.8%	1.33
6[a]	**0.000056**	**96.5 to 103.6%**	**1.00**
8	0.000040	97.1 to 103.0%	0.83
10	0.000031	97.5 to 102.6%	0.73
12	0.000025	97.7 to 102.4%	0.65
14	0.000021	97.9 to 102.2%	0.60
16	0.000019	98.1 to 102.0%	0.55

Note: Conclusions were reached using the input data of Table 11.2. These conclusions relate to a 3 + 3 assay having a dose ratio of 2:1.

[a] In routine work, typically, six plates are used. For this reason, six has been taken as the reference point for comparison.

In this three-dose levelthree-dose level assay, the ratio of adjacent dose levels is almost invariably 2:1. Therefore, putting the dose ratio as 2:1, the value of S_{xx} is calculated (as in the consideration of Example 9.2 of Chapter 9) thus:

$$S_{xx} = 18[(0.30103)^2 + (-0.30103)^2] = 3.262286$$

Putting the same arbitrary values for error squares and b as before (error squares = 0.8 and b = 8), we get $s^2 = 0.0320$ as before for six plates, and by Equation (11.2)

$$A = 0.0320/8^2 = 0.0005$$

then by Equation (11.3), first putting $M = 0$, we get

$$B = \left[\frac{1}{18} + \frac{1}{18} + \frac{0^2}{3.262286} \right] = 0.111111$$

The calculation of width of confidence limits is then identical with that shown in "The Effect of Replication" section earlier in this chapter. That is, the limits are 96.53 to 103.60% ($P = 0.95$), a range of 7.07%.

Now, put M at ± 0.05 corresponding to unknown potencies of 89 and 112% of standard. The value of B becomes

$$B = \left[\frac{1}{18} + \frac{1}{18} + \frac{+0.05^2}{3.262286} \right] = 0.111877$$

$$V(M) = A \times B = 0.0005 \times 0.111877 = 0.00005594$$

$$s_M = [0.00005594]^{0.5} = 0.0074792$$

Log percent confidence limits are $2 \pm 2.060 \times 0.0074792$ or 1.98459 and 2.01541, corresponding to confidence limits of 96.51 and 103.61%, a range of 7.10%. This range is only 0.4% greater than when the potency of the unknown was equal to that of the standard. However, as potencies of the unknown diverge further from that of standard, the range increases more rapidly. For example, when $M = \pm 0.301$, corresponding to 50 and 200% of standard, the range of confidence limits increases to 111.8% of that when $M = 0$.

Similar calculations lead to the figures presented in Table 11.4 for this three-dose level assay.

The Advantage of Symmetry

The foregoing examples have all been of symmetrical assays; that is, the number of dose levels and replications for the standard and each unknown have been identical.

Sometimes, an asymmetrical assay may appear to be the easiest option. For example, in a laboratory where the routine testing is by a two-dose level assay using

TABLE 11.4
The Variation in Width of Confidence Limits
Dependent on the Relative Potencies, Unknown vs.
Reference Standard

Relative Potencies	Confidence Limits (P = 0.95)	Range	Relative Range
25.0 or 400.0%	95.12 to 105.13%	10.01%	1.4158
50.0 or 200.0%	96.12 to 104.03%	7.91%	1.1188
80.0 or 125.0%	96.48 to 103.65%	7.17%	1.0141
90.0 or 111.1%	96.52 to 103.61%	7.09%	1.0028
95.0 or 105.3%	96.52 to 103.60%	7.08%	1.0014
100.0%	96.53 to 103.60%	7.07%	1.0000

Note: These values are based on a three-dose level symmetrical assay
having a dose ratio of 2:1.

petri dishes, the norm would be to compare two unknowns with the standard. On an occasion where only one unknown was to be assayed, the unknown test solutions may be applied to each plate twice so as to fully use its capacity for six zones. Thus, for the six-plate-unit assay consisting of 36 responses, the standard test solutions would have a replication of six, whereas that of the unknowns would be twelve.

Thus the value of B by Equation (11.4) would be:

$$B = [(1/12) + (1/24)] = 3/24 = 0.125000$$

Compare this with the corresponding three-dose level symmetrical assays for which

$$B = [(1/18) + (1/18)] = 1/9 = 0.111111$$

Giving A Equation (11.2) as before, the arbitrary value of $0.05/8^2 = 0.00078125$

$V(M)$ would become for the two-dose level and three-dose level assays, respectively:

$$0.00078125 \times 0.125000 = 0.000097656$$

and

$$0.00078125 \times 0.111111 = 0.000086805$$

The corresponding values for s_M would be 0.009882117 and 0.009316949, respectively.

A further small influence would arise from the number of degrees of freedom and consequent value of t. These would be as in the tabulation

Assay Design	d.f.	t
Two-dose level asymmetrical	27	2.052
Three-dose level symmetrical	25	2.060

Thus, the log confidence limits for the two cases would be, respectively,

$$2 \pm 2.052 \times 0.009882117 = 1.979722 \text{ to } 2.020278$$

and

$$2 \pm 2.060 \times 0.009316949 = 1.980807 \text{ to } 2.019193$$

and the percent confidence limits would be, respectively,

95.44 to 104.78%
95.68 to 104.52%

Thus, use of the asymmetrical assay would result in confidence limits of width 9.34% compared with 8.84% for the symmetrical assay — an increase of 5.7%. Although this is not a serious increase, it is something to be aware of and avoided if practicable.

This illustrates the general case that symmetrical assays are always more efficient than asymmetrical assays in the sense of better precision for the same number of observed responses. The same principle applies to slope ratio assays.

BIAS DUE TO CURVATURE

Another part of the folklore of biological assays concerns curvature. It may be inferred reasonably from the pharmacopoeial tests for curvature that significant curvature means that the assay is invalid.

In the antibiotic agar diffusion assay, the theory of zone formation suggests that the square of the zone width is directly proportional to the logarithm of antibiotic concentration. There is much practical evidence to confirm this is usually the case. If this is so, the zone diameter (unsquared) cannot also be directly proportional to the logarithm of antibiotic concentration. Nevertheless, over a short range of doses, the zone diameter-log dose line is near enough to being straight as to provide a workable basis for the calculation of potency estimate. In the days before computers, use of this approximation reduced the labor of calculation very substantially, and so it became the established procedure. It still remains the established procedure in laboratories throughout the world, even though it is a very simple matter to program a computer to do the longer but more rational calculation.

In assays having more than two-dose levels, the statistical evaluation includes a test for curvature. If curvature is found to be significant at the 5% level, the analyst is alerted to probable invalidity of the assay. Paradoxically, the more precise the assay, the more likely is curvature to be found statistically significant.

It was demonstrated by Hewitt (1981) that statistically significant curvature is frequently of little practical significance. The bias due to curvature may be quite small when compared with the calculated confidence limits of the potency estimate.

Figure 11.1 shows a plot of zone diameter versus log dose for a standard curve prepared in the investigation of a streptomycin assay using *Bacillus subtilis*. The eight dose levels having a ratio of 3:2, cover an overall range of about 17:1, which far greater than used in any normal assay. Curvature is readily apparent. The zone ꞈiameters were converted to zone width by subtracting the diameter of the reservoir (9 mm) and dividing by two. Figure 11.2 is a plot of square of zone width against log dose that clearly demonstrates the straight-line relationship in this case.

To investigate the extent of bias, the following series of calculations was made:

1. The slope of the straight line (Figure 11.2) was calculated as:

$$w^2 = 81.3417 + 7.0061x \tag{11.5}$$

where

$\quad w^2$ = square of zone width
$\quad x$ = coded log dose
81.3417 = the value of w^2 at the midpoint of the line

2. Using Equation (11.5) the "on the straight line" values of w^2 corresponding to the eight dose levels were calculated.

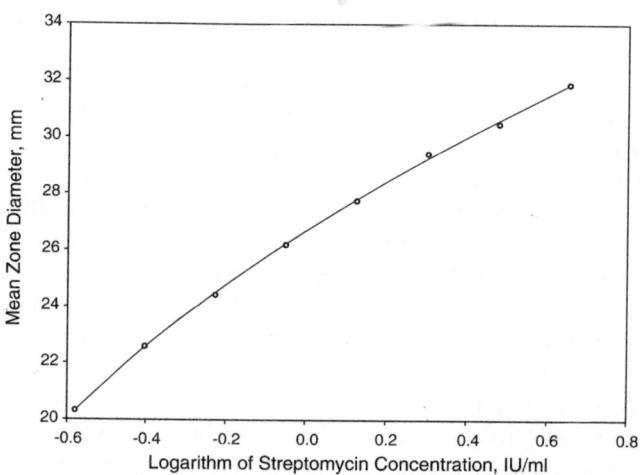

FIGURE 11.1 Responses to a series of dose levels of streptomycin standard in an agar diffusion assay using *Bacillus subtilis*. The mean of eight zone diameters for each dose level is plotted against the logarithm of dose. The curvature of the line is evident when viewed over the entire dose range (17:1) but not so evident over shorter ranges such as 4:1 or 2:1.

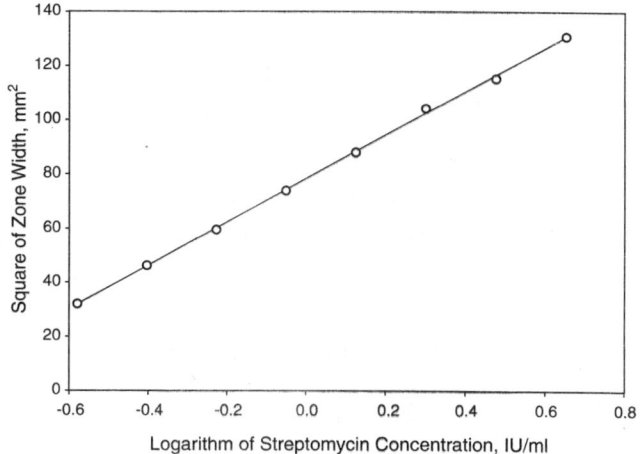

FIGURE 11.2 Responses to a series of dose levels of streptomycin standard in an agar diffusion assay using *Bacillus subtilis*. The raw data are the same as used in Figure 11.1, but zone widths were calculated as (zone diameter – 9)/2, where 9 is the diameter of the reservoir in mm. The square of zone width plotted against the logarithm of dose gave a straight line over the entire dose range.

3. From the eight "on the straight line" values, eight corresponding "on the curved line" values for zone diameter were calculated simply by taking the square root, doubling, and adding the diameter of the reservoir.
4. From the eight "on the curved line" values, an equation for the line was derived by means of orthogonal polynomial coefficients as

$$y = 27.037 + 8.899x - 2.156x^2 + 1.048x^3 - 0.935x^4 + 0.679x^5 \qquad (11.6)$$

5. Using Equation (11.6), it was possible to calculate the ideal responses (zone diameters free from random error) for any postulated dose.
6. Doses were postulated for standards and unknowns in various theoretical assays, and the responses then substituted in the appropriate expressions for calculation of potency estimate. Comparison of the potency estimate with the known potency (postulated potency of the unknown) gave a measure of the bias due to curvature in each individual case.

An illustration of such a calculation of an estimate of bias is given here by an example of a 3 + 3 assay. The following potencies were postulated:

Standard test solution potencies 0.5, 1, and 2 IU/ml
Unknown test solution potencies 0.315, 0.63, and 1.26 IU/ml

that is, exactly 63% of standard potencies.

Using Equation (11.6), zone diameters were calculated as

S_H	S_M	S_L	U_H	U_M	U_L
29.543	27.037	24.125	27.909	25.154	21.817

These figures were then substituted in the standard expressions for calculation of E and F obtained from the tabulation of Appendix 1. Thence a potency estimate was calculated, thus:

$$E = (1/4) \times [(27.909 + 29.543) - (21.817 + 24.125)] = 2.8775$$

$$F = (1/3) \times [(27.909 + 25.154 + 21.817) - (29.543 + 27.037 + 24.125)] = -1.9417$$

$$b = E/I = 2.84625/0.30103 = 9.5588$$

$$M = F/b = -1.975/9.45504 = -0.203128$$

$$R = \text{antilog } -0.203128 = 0.62643$$

The calculated potency ratio 0.62643 is 99.43% of the known value 0.63000, thus showing a bias of –0.57%.

Similar calculations were carried out for different assigned potencies of unknown and for different assay designs. The different assay designs examined in this way were:

2 + 2 with 2:1 dose ratio
2 + 2 with 4:1 dose ratio
3 + 3 with 2:1 dose ratio
5 + 1 with 5:4 dose ratio

The results of such calculations are displayed graphically in Figure 11.3 to Figure 11.6, based on a streptomycin standard curve. It will be seen that there is minimum bias in the case of the two-dose level assay with 2:1 dose ratio. The three-dose level assay succeeds in enhancing the unwanted effect of curvature, which it is designed to detect. The 5 + 1 assay shows greatest bias; it is always negative.

These calculations have their basis in a practically determined standard curve for streptomycin. Very similar patterns of bias were found in the cases of assays of tetracycline, ampicillin, and phenoxymethylpenicillin.

As a part of the same study, it was shown by Hewitt (1981) that a quadratic component of curvature is entirely without influence on the estimated potency in a *symmetrical parallel-line* assay. The algebraic proof of this is given in Appendix 8.

Quadratic curvature *does* bias the potency estimate in asymmetrical assays and higher degrees of curvature influence the result in all assays. It is stressed that these findings refer only to parallel-line assays.

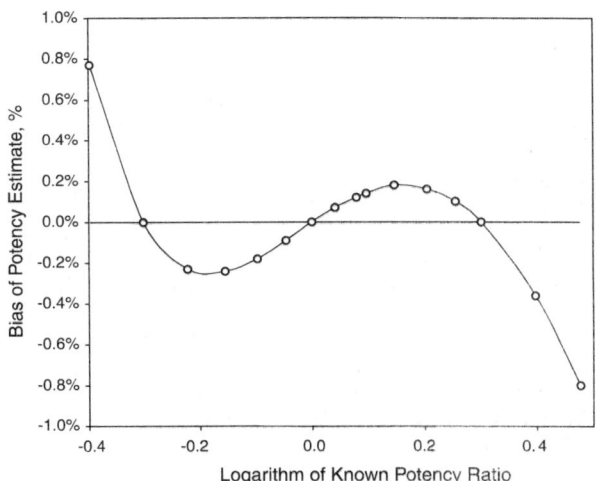

FIGURE 11.3 Illustration of the variation of bias due to curvature with changing true potency ratio of unknown to reference standard in a streptomycin assay using *Bacillus subtilis* in a 2 + 2 assay with 2:1 dose ratio.

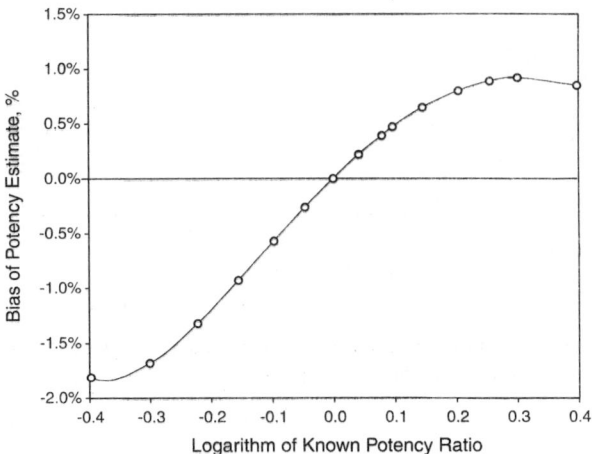

FIGURE 11.4 Illustration of the variation of bias due to curvature with changing true potency ratio of unknown to reference standard in a streptomycin assay using *Bacillus subtilis* in a 2 + 2 assay with 4:1 dose ratio.

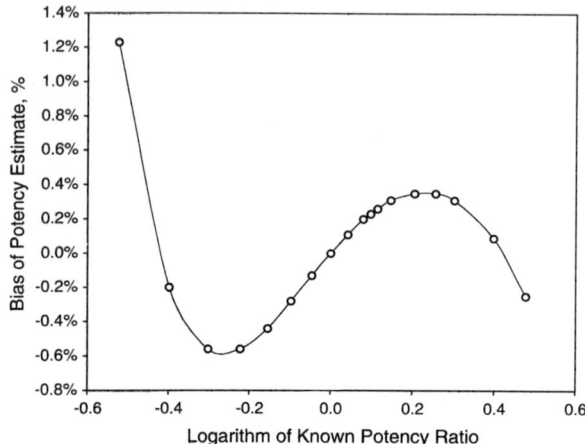

FIGURE 11.5 Illustration of the variation of bias due to curvature with changing true potency ratio of unknown to reference standard in a streptomycin assay using *Bacillus subtilis* in a 3 + 3 assay with 2:1 dose ratio.

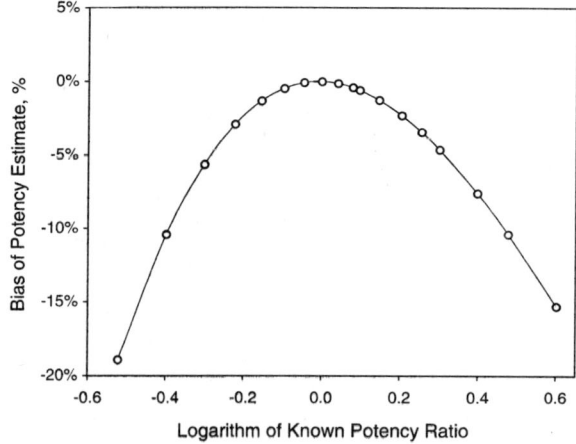

FIGURE 11.6 Illustration of the variation of bias due to curvature with changing true potency ratio of unknown to reference standard in a streptomycin assay using *Bacillus subtilis* in a 5 + 1 assay with 5:4 dose ratio.

NONPARALLELISM DUE TO CURVATURE

Naturally, failure of response lines to be parallel in a parallel-line assay is cause for concern. Yet even this is not necessarily a disaster. As the zone diameter vs. log dose line is curved, if an unknown potency is not close to that of standard, it will appear that the lines are not parallel. Such a case does not represent an invalid assay. This apparent problem may be resolved by calculating squares of zone widths from the zone diameters before embarking on the standard calculation. In conclusion, the results of statistical evaluation have to be interpreted thoughtfully.

CHOOSING A DESIGN FOR A TURBIDIMETRIC GROWTH-INHIBITING SUBSTANCE ASSAY

The *British Pharmacopoeia* (*BP*) (1993), Appendix XIVA, A167 offers the following guidance:

> Rectilinearity of the dose-response relationship, transformed or untransformed, is often obtained over a very limited range. It is this range that must be used in calculating the activity and it must include at least three consecutive doses in order to permit rectilinearity to be verified.

Although the statement is "at least three," the pharmacopoeia recognizes that more than three levels is unlikely to be attained due to the limited range of rectilinearity, as is shown by the footnote:

> In order to obtain the required rectilinearity it may be necessary to select from a larger number, three consecutive doses, using corresponding doses of the standard preparation and of the substance being examined.

It is because of the narrow range of rectilinearity that turbidimetric growth-inhibiting substance assays generally employ a low dose ratio such as 3:2 or 4:3, thus minimizing the overall range.

Compare two possible designs employing 12 tubes per preparation:

1. A three-dose level assay having a 3:2 ratio; there would be four tubes for each dose level and the relative doses would be:
   1:1.5:2.25

2. A four-dose level assay having a 4:3 ratio; there would be three tubes for each dose level, and the relative doses would be approximately
   1:1.33:1.78:2.37

The two designs have a similar overall dose range. In accordance with the principles explained in "The Effect of the Value of M on Width of Confidence Limits" earlier in this chapter, design 2 will yield potency estimates that are slightly less precise than those of design 1 when estimated potency ratio departs from 1.00. This potential loss in precision is not a big problem; however, it is a problem that has been introduced at the cost of greater practical effort in preparing the additional dilutions.

If the dilution process is completely automated, then the greater practical effort is of no consequence. A possible advantage of using four-dose levels is that, in the event of nonrectilinearity over the entire range, it will be possible to select and use only three levels for which the log dose-response relationship is near enough to being rectilinear to meet statistical requirements and so achieve a valid assay.

A possible arrangement for tubes in a rack for a four-dose level assay with a replication of four is:

0	S_1	S_2	S_3	S_4	T_1	T_2	T_3	T_4
0	S_1	S_2	S_3	S_4	T_1	T_2	T_3	T_4
0	S_1	S_2	S_3	S_4	T_1	T_2	T_3	T_4
0	S_1	S_2	S_3	S_4	T_1	T_2	T_3	T_4

Although ideally, the tubes should be placed in the rack in accordance with a randomized pattern, this may cause operational problems that are time consuming and accident prone. Problems of positional differences are perhaps better resolved by the use of a very well-stirred water bath. Kavanagh (1972) describes a high-precision water bath used in the Autoturb™ system. It is claimed that temperatures in different positions in the bath do not differ by more than 0.02°C.

CHOOSING A DESIGN FOR TURBIDIMETRIC ASSAY OF A GROWTH-PROMOTING SUBSTANCE

Although the possible experimental designs for tube assays are restricted by the capacity of the tube rack, the possibilities are numerous; suffice it to say that each rack, in which all tubes will be subjected to the same conditions of incubation, should contain one tube representing every treatment in the assay block. For example, a rack with a capacity for 30 tubes could accommodate tubes representing the standard and two unknowns of a symmetrical seven-point common-zero assay as follows:

0	S_1	S_2	S_3	T_1	T_2	T_3	T'_1	T'_2	T'_3
0	S_1	S_2	S_3	T_1	T_2	T_3	T'_1	T'_2	T'_3
0	S_1	S_2	S_3	T_1	T_2	T_3	T'_1	T'_2	T'_3

As in the case of turbidimetric growth-inhibiting substance assays, ideally, the tubes should be placed in the rack in accordance with a randomized pattern. Vincent (1989) describes suitable designs. However, for the reasons explained in Chapter 3, control of temperature is less critical provided that incubation time is long enough.

Using methods analogous to those used for parallel-line assays, similar conclusions were reached concerning replication and the superiority of symmetrical designs.

Consider now whether multiple-dose-level assays have any advantage over the two-dose level, five-point common-zero assay. If the larger number of dose levels is attained by increasing the overall range of doses, this increases the possibility of getting into the nonlinear range of responses.

If the larger number of dose levels is attained by keeping the overall range the same, that problem does not arise. However, such assays involve more work in preparing dilutions. It can be shown by reasoning analogous to that used in "The Effect of the Value of M on Width of Confidence Limits" that assays having more than two-dose levels are somewhat less precise than the five-point common-zero assay. Thus, more work will result in a less precise assay.

It is perhaps for these reasons that only the two-dose level assay was described in the second and third editions of the *European Pharmacopoeia* (1993, 1997). It is not clear why three- and four-dose level assays were introduced in the *European Pharmacopoeia* 2000. The five-point common-zero assay is the best design to use for the assay of a pharmaceutical sample for which a nominal potency can be assigned with confidence.

Asymmetric assays are sometimes used; for example, an assay may have three or more dose levels of standard and one or two of unknown. Such assays are useful when the potency of the unknown cannot be guessed with any confidence before the assay. Potency may be estimated by interpolation from the standard curve. If an accurate potency estimate is required, then the initial result by interpolation may be used as a starting point for a further assay using a symmetrical design.

The effect of replication on width of confidence limits for five-point and seven-point assays is illustrated in Table 11.5. The values in the table were obtained by calculations analogous to those for parallel-line assays decribed in "Factors Influencing Width of Confidence Levels" earlier in this chapter.

In contrast to parallel-line assays, no theoretical features can be offered here to quantify the influence of curvature. The influence of curvature may be illustrated simply by reference to two practically determined dose-response lines for (1) nicotinic acid and (2) thiamine. For these two dose-response lines, equations were fitted to the curves relating dose to response. Then "responses" could be calculated to assumed doses of standard and unknowns.

TABLE 11.5
The Change in Width of Confidence Limits in Slope Ratio Assays According to Degree of Replication

Replication	Relative Width of Confidence Limits for a Five-Point Assay	Relative Width of Confidence Limits for a Seven-Point Assay
2	1.41	1.35
3	**1.00**	**1.00**
4	0.83	0.84
5	0.72	0.74
6	0.65	0.68

Note: The relative value for a replication of three is arbitrarily set at unity.

TABLE 11.6

Two Illustrations of Bias in the Calculated Potency Estimate Due to Curvature of the Response Line in Slope Ratio Assays

	Nicotinic Acid		Thiamine	
True Potency Ratio	**Calculated Ratio**	**Bias, %**	**Calculated Ratio**	**Bias, %**
0.50	0.524	+4.7	0.632	+26.4
0.80	0.817	+2.1	0.873	+9.1
1.00	1.000	0.0	1.000	0.0
1.25	1.212	−3.0	1.134	−9.2
1.50	1.348	−9.4	1.251	−16.6

The two equations were, respectively,

$$y = 35.936 + 9.271x - 0.413x^2 - 0.021x^3 \qquad (11.7)$$

$$y = 48.300 + 7.865x - 1.377x^2 + 0.288x^3 - 0.020x^4 - 0.003x^5 \qquad (11.8)$$

For example, in the case of nicotinic acid, doses were assumed to be 1.0 and 2.0 for the standard and 0.8 and 1.6 for the unknown. Putting these values in Equation (11.7), responses are calculated. Then, putting the calculated "responses" into the standard formula for calculating potency estimate for a five-point assay, the unknown potency was calculated as 81.7% of the standard, as compared with the known value, 80.0%. Thus the bias was (81.7 × 100)/80.0 = 2.1%. Values calculated in this way are shown in Table 11.6 for two vitamins.

The standard-response lines and calculated-unknown-response lines for these two assays are shown in Figure 11.7 and Figure 11.8. It is clear that when visual inspection shows obvious curvature, the slope ratio calculation should not be used.

SUMMARY AND CONCLUSIONS

Some general comments are equally applicable to parallel-line and slope ratio assays.

Symmetrical assays are inherently more efficient than asymmetrical assays. *Most efficient* means *highest precision for least effort.* When more than one unknown is to be tested, multiple symmetrical assays may be the most efficient. In these, each unknown is represented in the assay block with the same replication and same number of dose levels as the standard.

Unless automation is such that practical effort is of no consequence, a smaller number of dose levels is more efficient than a higher number, in that less effort is made in preparing the test solutions. A smaller number of dose levels generally means a lower overall dose range, which minimizes curvature of the response lines.

Increasing replication increases precision and, as a rough guide, a fourfold increase in replication results in halving the width of the confidence limits.

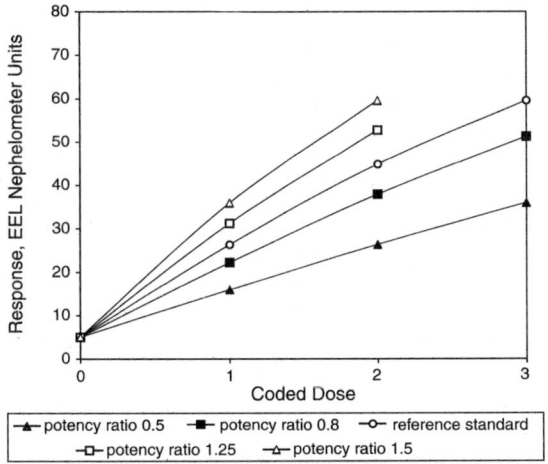

FIGURE 11.7 Dose-response lines for a nicotinic acid assay. The standard line is a practically determined line. Lines representing 0.5, 0.8, 1.25, and 1.5 times the standard potency were obtained by calculating values from the equation for the standard curve.

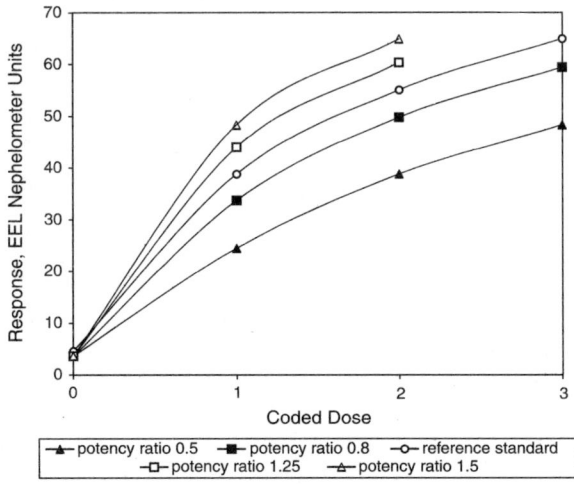

FIGURE 11.8 Dose-response lines for a thiamine assay. The standard line is a practically determined line. Lines representing 0.5, 0.8, 1.25, and 1.5 times the standard potency were obtained by calculating values from the equation for the standard curve.

REFERENCES

Dabbah, R. 1994. Personal communication from USP Convention.

Dabbah, R. 1995. Personal communication from USP Convention.

Hewitt, W. 1977a. *Microbiological Assay: An Introduction to Quantitative Principles and Evaluation*, San Diego: Academic Press, p. 50.

Hewitt, W. 1977b. *Microbiological Assay: An Introduction to Quantitative Principles and Evaluation*, San Diego: Academic Press, p. 57.

Hewitt, W. 1981. Influence of curvature of response lines in antibiotic agar diffusion assays, *J. Biol. Standardisation*, 9, 1.

Kavanagh, F.W. 1972. *Analytical Microbiology*, Vol II, San Diego: Academic Press.

Lees, K.A. and Tootill, J.P.R. 1955a. Microbial assay on large plates, part I, *Analyst,* 80, 95.

Lees, K.A. and Tootill, J.P.R. 1955b. Microbial assay on large plates, part II, *Analyst,* 80, 531.

The European Pharmacopoeia. 1993. Second Edition, Strasbourg: Council of Europe.

The European Pharmacopoeia. 1997. Third Edition, Strasbourg: Council of Europe.

The European Pharmacopoeia. 2000. Third Edition, Strasbourg: Council of Europe.

The European Pharmacopoeia. 2001. Third Edition Supplement, Strasbourg: Council of Europe.

The United States Pharmacopoeia 24. 2000. <111> Design and analysis of biological assays.

The British Pharmacopoeia. 1993. Appendix XIV A.

Vincent, S. 1989. *Theory and Application of Microbiological Assay*, edited by W. Hewitt and S. Vincent, San Diego: Academic Press.

12 Concluding Thoughts

INTRODUCTION TO THE END

In the preceding chapters, we have considered the theory and practice of the three main types of microbiological assay. We have considered the objectives of the assay, the degree of accuracy, precision and specificity required and discussed choice of assay design appropriate to the objectives of the assay. Finally, we need to decide how we shall report the results of our work. Normally, assays are done at least in duplicate but sometimes the replication may be much higher, for example in the calibration of a new reference standard. A simple mean of all the potency estimates is not the best estimate of the true mean. In subsequent sections procedures will be described leading to a weighted mean together with its confidence limits.

COMBINATION OF RESULTS FROM REPLICATE ASSAYS

In any assay, be it chemical, physicochemical, or biological, it is common practice, and a desirable practice, to carry out at least two assays on any unknown. Intuitively, if agreement between two or more potency estimates is close, the analyst feels confidence in the results and the average will seem to be nearer to the truth than any individual estimate.

Usually, the individual estimates are not of equally good precision, so that a simple mean will not be the best estimate from the available data.

Whatever procedure is used for calculating a mean, all estimates of potency must be corrected for the potency assigned to each preparation before they can be combined. That is, the estimated potencies must be adjusted to eliminate any variation attributable to difference in concentration in the test solutions. Calculations will, therefore, be based on adjusted values of M, the logarithm of the potency estimate.

Pharmacopoeias give guidance on procedures for combining individual estimates to give a mean together with its confidence limits. Reference will be made here to the guidance of the *European, International*, and *United States Pharmacopoeias*.

As the procedures refer to *independent* estimates, it is important to understand what is meant by *independent*.

The *European Pharmacopoeia* (2001, 6.1, 32) states:

Two assays may be regarded as mutually independent when the execution of either does not affect the probabilities of the possible outcomes of the other. This implies that the random errors in all essential factors influencing the result (for example, dilutions of the standard and the preparation to be examined, the sensitivity of the biological indicator) in one assay must be independent of the corresponding random errors in the other one. Assays on successive days using the original and retained dilutions of the standard therefore are not independent assays.

This definition appears to pose some problems in the case of a microbiological assay using a spore suspension. As the same spore suspension is intended to be used over many months, does this mean that all assays using that suspension are not independent? If a change in spore suspension were to cause a change in response, then certainly assays using a single spore suspension would not be independent.

The known facts are that if standard and unknown are qualitatively identical, then a change of actual test organism, although quite possibly making a difference to the precision of the potency estimate, would not cause any bias. Thus it is reasoned that when standard and unknown are known to be qualitatively identical, a series of assays using the same spore suspension could be regarded as independent.

However, in the case of an antibiotic such as neomycin or erythromycin, where standard and unknown consist of the same related substances, the situation could be different. Two or more assays using a single spore suspension could be regarded as independent only if there were evidence that either (1) the proportions of the different components were the same in standard and unknown preparations or (2) change of spore suspension did not lead to any bias.

To summarize, it is suggested that use of the same spore suspension for two or more assays would not necessarily be cause to classify those assays as non-independent.

CALCULATION OF A WEIGHTED MEAN BASED ON INTERNAL ASSAY VARIATION

The simplest procedure will be described now. It is that of the *International Pharmacopoeia* (1979). (The current version of the *IP* relating to microbiological assay of antibiotics is Volume 1 of the third edition [1979]. However, for calculation of results, the reader is referred to the second edition [1967], Appendix 45, Biological Assays and Tests.)

For each individual assay, a statistical weight is calculated. Using the approximate method of calculating confidence limits for an assay, an intermediate step is the calculation of the variance of M, [$V(M)$ Equation (9.23)]. The statistical weight, W, is simply calculated as

$$W = 1/V(M) \qquad (12.1)$$

The weighted mean is then calculated by forming the products WM for each assay. The sum of these products divided by the sum of the individual weights then gives the logarithm of the weighted mean potency, $\bar{M}$

$$\bar{M} = \frac{\Sigma WM}{\Sigma W} \qquad (12.2)$$

The standard error of the logarithm of weighted mean potency is obtained as the square root of the reciprocal of the total weight thus:

$$s_{\overline{M}} = [1/\Sigma W]^{0.5} \tag{12.3}$$

Confidence limits of the logarithm of mean potency estimate are calculated as:

$$\overline{M} + ts_{\overline{M}} \tag{12.4}$$

where the number of degrees of freedom for t is the total of the degrees of freedom for all the assays.

The methods described in the *European, International* and *United States Pharmacopoeias* differ superficially; different symbols are used and some procedures appear much more complex than others. It will be demonstrated that they are essentially similar.

EUROPEAN PHARMACOPOEIA GUIDANCE

Preliminary tests must be carried out to determine whether the calculation of the weighted mean will be valid.

The *EP (2000)* lists four conditions that must be fulfilled if the method is to be used:

1. The potency estimates are derived from independent assays.
2. For each assay, C is close to 1 (say, less than 1.1); alternatively, this may be expressed as g is less than 0.1. Here, it is important to recognize the special case of *micro*biological assay for which g is almost invariably not only less than 0.1 but less than 0.01.
3. The number of degrees of freedom of the individual residual error is not smaller than 6, but preferably larger than 15.
4. The individual potency estimates form a homogenous set.

Condition 4 means that one must *first* calculate the weighted mean and *then* check that the calculation has led to a valid result.

For calculation of statistical weight, W, the *European Pharmacopoeia* (2000) gives the expression:

$$W = \frac{4t^2}{L^2} \tag{12.5}$$

in which

t is Student's t dependent on the number of degrees of freedom in the test
L is the difference between the logarithms of the upper and lower confidence limits of the estimated potency

However, when $V(M)$ is calculated routinely in the assay evaluation procedure by Equation (9.21), weights may be calculated more simply as

$$W = 1/V(M) \qquad (12.6)$$

It is important to recognize that the numerical value of the weights depends on whether ln or $\log_{10}$ is used in the calculation — either may be used, but consistency is essential. Also, whether Equation (12.5) or Equation (12.6) is used, it is necessary to use a large number of decimal places.

The weights thus calculated are then used as described in "Combination of Results from Replicate Assays" earlier in this chapter to calculate the weighted mean potency and confidence limits of that mean using Equation (12.2) to Equation (12.4).

Now we return to the question of homogeneity of the potency estimates to establish whether the weighted mean potency and confidence limits as calculated above have a valid basis. The pharmacopoeia directs that a statistic be calculated that is distributed approximately as X^2. It is calculated as

$$X^2 = \Sigma W \left(M - \overline{M} \right)^2 \qquad (12.7)$$

If the calculated value of X^2 is smaller than the tabulated value for the appropriate number of degrees of freedom, the mean potency and confidence limits will be meaningful. (See Table 8.3 of the pharmacopoeia or Appendix 9 of this book.)

INTERNATIONAL PHARMACOPOEIA GUIDANCE

The *IP* (1967) states that the most accurate methods of combining results of repeated assays are complicated. It describes the method (which we have seen at the beginning of this chapter) that may be used when g is less than 0.1. This corresponds with the *European Pharmacopoeia* statement that C should be less than 1.112 since $C = 1/(1 - g)$.

UNITED STATES PHARMACOPOEIA GUIDANCE

The *USP 24* directs that the separate estimates of M, the logarithm of potency estimate, be tested for mutual consistency. If the Ms are consistent, their respective confidence intervals will overlap. If confidence intervals do not overlap, or if overlap is small, a further check must be carried out as described now.

For each of the individual assays calculate a weight. The *USP* expression for calculating weight is identical with that of the *EP*, that is, Equation (12.5).

From the weights, calculate an approximate X^2 with (h – 1) degrees of freedom as

$$\text{Approx. } X^2_M = \Sigma \left(W M^2 \right) - \left[\Sigma (WM) \right]^2 / \Sigma W \qquad (12.8)$$

The approximate value of X_M^2 is then compared with tabulated critical values for X^2 (Table 9 of <111> of the pharmacopoeia or Appendix 9 of this book). The pharmacopoeia then directs that if the approximate value of X_M^2 is well under the tabulated value the weights may be used to calculate a mean weighted value of M using Equation (12.2) and confidence limits. If the calculated value of X_M^2 approaches or exceeds the tabulated critical value, the pharmacopoeia offers alternative procedures.

Again, the special case of *micro*biological assays is cited. It is suggested that in a well-conducted *micro*biological assay laboratory, replicate assays will be mutually consistent, and so the alternative procedures are not discussed here. If that is not the case, then the reason should be sought by reviewing practical operations.

COMPARISON OF THE *USP* AND *EP* PROCEDURES FOR TESTING FOR CONSISTENCY OF POTENCY ESTIMATES

Both pharmacopoeias describe procedures for calculation of an approximate value for X^2. Although these both lead to the same value, the recommended interpretation procedures differ. The *EP* states simply that if the calculated value of X^2 is less than the tabulated critical value, then the values calculated for mean potency and confidence limits will be meaningful. In contrast, the *USP* requires that the calculated value of X^2 be well under the tabulated critical value.

COMPARISON OF THE *EUROPEAN* AND *INTERNATIONAL* PHARMACOPOEIAL METHODS

The relationship between the two formulas for statistical weight may be seen easily by writing them both in approximated form. First, the full versions are shown:

1. The method of the *IP*, the traditional method.
   Weight, $W = 1/V(M)$, where the variance of M, $V(M)$ is as previously shown in Chapter 9 [Equation (9.23)], which is reproduced here for convenience.

$$V(M) = \frac{s^2}{b^2}\left[\frac{1}{N_s} + \frac{1}{N_u} + \frac{M^2}{S_{xx}}\right] \tag{9.23}$$

2. The method of the *EP* (*EP* formula).

$$W = \frac{b^2}{s^2 C}\left[\frac{1}{\dfrac{1}{N_s} + \dfrac{1}{N_u} + \dfrac{(y_s - y_u)^2}{E - s^2 t^2}}\right] \tag{12.9}$$

3. An approximated form of the traditional formula.
    Simplifying 1, we can say $(M^2)/S_{xx}$ is always small (except in a *very* poor assay). So we can write the approximate expression:

$$V(M) = \frac{s^2}{b^2} \left[\frac{1}{N_s} + \frac{1}{N_u} \right]$$ (12.10)

4. An approximated form of the *EP* formula.
    Now simplifying 2, we can say that C is always very close to 1.00 (except in a *very* poor assay); also that in the part of the expression

$$\frac{\left(y_s - y_u\right)^2}{E - s^2 t^2}$$

E is always relatively large, and the numerator is always relatively small (approaching zero as sample and standard test solution potency ratio approaches 1.00), thus this part of the expression may be omitted. Now rewriting with these modifications we get

$$W = \frac{b^2}{s^2} \left[\frac{1}{\dfrac{1}{N_s} + \dfrac{1}{N_u}} \right]$$ (12.11)

Writing the reciprocals of both sides of this expression we have:

$$1/W = \frac{s^2}{b^2} \left[\frac{1}{N_s} + \frac{1}{N_u} \right]$$ (12.12)

We know from Equation (12.1) that $1/W = V(M)$. Thus, the essential similarity has been demonstrated.

However, the two methods may give quite different values for weights. The reason that the figures obtained for weights may be different is that the traditional method uses logarithms to base 10 whereas the *EP* suggests the use of natural logarihms. The value of the slope b is dependent on which logarithm is used, hence the different values for weight.

In a comparison of the two procedures using the same raw data from several assays, it was found that with either method the *relative* weights from the individual assays were identical and, thus, the weighted mean potencies calculated by both methods were identical.

APPLICATION OF THE VARIOUS FORMULAS TO SOME PRACTICAL RESULTS

The calculation of a weighted mean is illustrated in Example 12.1, which follows.

EXAMPLE 12.1 THE EXAMINATION AND COMBINATION OF RESULTS OF FOUR PENICILLIN ASSAYS

The input data are presented in Table 12.1. The values of M in the table have been adjusted to eliminate variation due to concentration differences in the test solutions.

An approximate value of X^2 is calculated as illustrated in Table 12.2.

TABLE 12.1
The Essential Data from Four Independent Penicillin Assays

Assay Number	Estimated Potency IU/mg	Confidence Limits IU/mg ($P = 0.95$)	Adjusted M	$V(M)$	Weight W
1	1629	1570–1691	+0.0092	0.0000628	15,835
2	1613	1541–1688	+0.0048	0.0000930	10,749
3	1586	1530–1644	−0.0025	0.0000578	17,625
4	1558	1485–1635	−0.0103	0.0001050	9,524

TABLE 12.2
The Essential Data for Calculating an Approximate Value of X^2 by the *USP* Procedure, Equation (12.7)

Assay Number	M	W	WM	M^2	WM^2
1	+0.0092	15,835	145.6820	0.00008464	1.340274
2	+0.0048	10,749	51.5952	0.00002304	0.247657
3	−0.0025	17,625	−44.0625	0.00000625	0.110156
4	−0.0103	9,524	−98.0972	0.00010609	1.010401
Totals		53,733	55.1175		2.708488

Approx X^2 = 2.708489 − (55.1175)²/53,733

= 2.708489 − 0.056538 = 2.651951

The approximate value of X^2 (2.652) is much less than the critical value for 3 d.f. (7.82) from the table of Appendix 9, and so the potency estimates may be combined using Equation (12.2).

$$\overline{M} = 55.1175/53,733 = 0.0010258$$

so that the weighted mean potency estimate is

antilog 0.0010258 = 1.0024 or 100.24%

The corresponding variance of the weighted mean of M is given by Equation (12.1) as

$$V\left(\overline{M}\right) = 1/53,733 = 0.000018561$$

and the corresponding standard error is obtained by Equation (9.24) as:

$$0.000018611^{0.5} = 0.004314$$

The error term in each of the four assays had 25 degrees of freedom so there were 100 d.f. for the combined assay. The corresponding value of t is 1.984, thus the log percent confidence limits are obtained by Equation (9.25) as:

$$2 \pm 1.984 \times 0.00434$$

and percent confidence limits as

98.05 to 101.99% ($P = 0.95$)

CURRENT TRENDS

A major feature of this book is the evaluation and interpretation of data arising from microbiological potency assays. The mathematical basis for evaluation of biological assays in general was investigated and recorded a little over half a century ago by such eminent scientists as Berkson, Bliss, Emmens, Finney, Fisher, and Yates, to whom references are made throughout this book. Nothing has changed these fundamental mathematical truths.

However, official guidance on evaluation does tend to vary with each new edition of a pharmacopoeia. It is important to realize that pharmacopoeial guidance refers to biological assays in general and does not recognize the special case of *microbiological* assays, in which the real source of error is physical rather than biological and their precision is comparable with that of physicochemical assay methods, such as gas or high-performance liquid chromatography.

Pharmacopoeial Changes

It is important to recognize that pharmacopoeial guidance is usually preceded by a statement such as:

Alternative methods may be used provided that they are not less reliable than those described here.

Such provisos apply equally in the European Union and the United States.

Over the past decade, there have been several changes in pharmacopoeial guidance. The logic of these changes is not immediately obvious to the analyst engaged in his or her day-to-day responsibilities, and so it is fortunate that observation of these changes is not mandatory.

Some noteworthy changes include:

1. For several decades the British, International, and United States pharmacopoieas used logarithms to base 10 in the examples given to illustrate calculation procedures. The earlier editions of the *European Pharmacopoeia* chose to use natural logarithms in its illustrative examples. In the year 2000 Supplement, however, it stated that logarithms to base 10 can be used equally well. That, of course, was always apparent.
2. Change of the meaning of symbols. One example is given here: In the *EP* (1997), the symbols S_1 to S_n represented the *total* response to differing doses of the standard preparation; in the 2000 Supplement of that pharmacopoeia the same symbols represent *mean* responses. Such a change was not helpful.
3. Recommendation that, with the common availability of computers, there is no longer any need for the use of the simpler computational (postanalysis of variance) formulas for calculation of confidence limits. Although this is true, it overlooks the fact that in *micro*biological assay, because of its intrinsically higher precision than *macro*biological assay, results obtained by the more complex calculation are likely to differ from those by the simpler calculation by a mere 0.02 percent. The simpler calculation, being easier to understand, has a greater educational value.

INTERPRETATION ON BASIS OF EXPERIENCE

An important feature of *European Pharmacopoeia* guidance on assay evaluation is involvement of the analyst in the decision-making process. This occurs at two stages of the evaluation procedure:

1. The raw data (unscrambled if necessary) should be inspected before proceeding to the statistical evaluation and calculation of potency estimate.
2. At stages following the analysis of variance, the analyst should assess whether the data is valid and suitable for further processing to arrive at a potency estimate with confidence limits. In other words, the analyst and *not* the computer should make the decision to accept or reject.

SECURITY

Arising from new 21CFR Part II regulatory requirements in the United States and in Europe (*Europoean Pharmacopoeia* 2001) there is now great awareness of the need to build security safeguards into computer programs. These requirements endeavor to block any actions intended to tamper with the program itself or with assay data that have been recorded.

Although efforts to counter intentional distortion of assay results are commendable, we should not underestimate the importance of unintentional distortion of results. The hazard of operator anticipation of zone size has been mentioned before. Based on experience of what is normally attainable in reproducibility of diameter in replicate zones of the same treatment, I have often seen results that I judge as being too good to be true. Such problems will not be resolved by more regulations. What is needed is a good perception of the potential problems and good management.

REFERENCES

International Pharmacopoeia. 1967.
International Pharmacopoeia. 1979.
The European Pharmacopoeia. 2000. 6, 6.1.
The European Pharmacopoeia. 2001. 6.1, 32.
The United States Pharmacopoeia 24.

Appendices

APPENDIX 1
Calculations for Parallel-Line Assays

Expressions for calculating the values of E and F according to number of dose levels.
The terms E and F were introduced in Chapter 2. Their meanings are illustrated graphically in Figure 2.7.

Number of Dose Levels	Expressions for the Calculation of E	Expressions for the Calculation of F
2	$E = [(S_2 + U_2) - (S_1 + U_1)]/2$	$F = [(U_2 + U_1) - (S_2 + S_1)]/2$
3	$E = [(S_3 + U_3) - (S_1 + U_1)]/4$	$F = [(U_3 + U_2 + U_1) - (S_3 + S_2 + S_1)]/3$
4	$E = [3(S_4 + U_4) + (S_3 + U_3) - (S_2 + U_2) - 3(S_1 + U_1)]/20$	$F = [(U_4 + U_3 + U_2 + U_1) - (S_4 + S_3 + S_2 + S_1)]/4$
5	$E = [2(S_5 + U_5) + (S_4 + U_4) - (S_2 + U_2) - 2(S_1 + U_1)]/20$	$F = [(U_5 + U_4 + U_3 + U_2 + U_1) - (S_5 + S_4 + S_3 + S_2 + S_1)]/5$

Note: S_1 is the mean response to treatment S_1. S_n and U_n have analogous meanings.

APPENDIX 2
Latin and Quasi-Latin Square Designs

(1) 6 × 6 squares
Suitable for:

Two samples and one standard each at two-dose levels.
One sample and one standard each at three-dose levels.

Design GF01

2	5	6	3	4	1
4	6	2	1	5	3
1	3	5	4	2	6
3	1	4	5	6	2
5	2	1	6	3	4
6	4	3	2	1	5

Design GF02

2	4	6	1	5	3
5	1	3	4	2	6
3	5	2	6	1	4
4	6	1	5	3	2
1	3	4	2	6	5
6	2	5	4	3	1

Design GF03

6	1	4	3	2	5
3	6	1	5	4	2
1	3	5	2	6	4
4	5	2	1	3	6
2	4	3	6	5	1
5	2	6	4	1	3

Design GF04

3	5	2	4	1	6
1	4	6	5	2	3
6	3	5	2	4	1
4	2	1	3	6	5
2	1	3	6	5	4
5	6	4	1	3	2

Design GF05

4	1	6	3	5	2
2	6	3	4	1	5
5	3	2	1	4	6
6	5	1	2	3	4
3	2	4	5	6	1
1	4	5	6	2	3

Design GF06

5	4	6	3	1	2
2	6	5	1	3	4
4	2	1	6	5	3
3	1	2	5	4	6
1	3	4	2	6	5
6	5	3	4	2	1

Design GF07

4	2	1	3	6	5
6	1	2	5	4	3
1	5	4	2	3	6
2	6	3	1	5	4
5	3	6	4	2	1
3	4	5	6	1	2

Design GF08

2	6	3	4	1	5
5	3	2	1	4	6
6	5	1	2	3	4
3	2	4	5	6	1
1	4	5	6	2	3
4	1	6	3	5	2

Design GF09

1	4	2	5	3	6
3	6	1	2	5	4
6	5	4	3	1	2
5	3	6	4	2	1
2	1	3	6	4	5
4	2	5	1	6	3

Design GF10

1	4	5	6	3	2
4	5	6	3	2	1
5	1	4	2	6	3
2	6	3	1	5	4
6	3	2	4	1	5
3	2	1	5	4	6

Design GF11

3	5	6	4	2	1
2	3	5	1	6	4
6	1	3	2	4	5
5	4	2	3	1	6
1	2	4	6	5	3
4	6	1	5	3	2

Design GF12

4	6	1	5	3	2
3	5	6	4	2	1
2	3	5	1	6	4
6	1	3	2	4	5
5	4	2	3	1	6
1	2	4	6	5	3

Design GF13

5	3	6	4	2	1
3	4	5	6	1	2
4	2	1	3	6	5
6	1	2	5	4	3
1	5	4	2	3	6
2	6	3	1	5	4

Design GF14

4	5	6	3	2	1
5	1	4	2	6	3
2	6	3	1	5	4
6	3	2	4	1	5
3	2	1	5	4	6
1	4	5	6	3	2

Design GF15

6	5	3	4	2	1
5	4	6	3	1	2
2	6	5	1	3	4
4	2	1	6	5	3
3	1	2	5	4	6
1	3	4	2	6	5

Design GF16

1	4	5	6	2	3
4	1	6	3	5	2
2	6	3	4	1	5
5	3	2	1	4	6
6	5	1	2	3	4
3	2	4	5	6	1

Design GF17

3	2	1	5	4	6
1	4	5	6	3	2
4	5	6	3	2	1
5	1	4	2	6	3
2	6	3	1	5	4
6	3	2	4	1	5

Design GF18

2	6	5	1	3	4
4	2	1	6	5	3
3	1	2	5	4	6
1	3	4	2	6	5
6	5	3	4	2	1
5	4	6	3	1	2

Design GF19

3	4	5	6	1	2
4	2	1	3	6	5
6	1	2	5	4	3
1	5	4	2	3	6
2	6	3	1	5	4
5	3	6	4	2	1

Design GF20

4	2	5	1	6	3
1	4	2	5	3	6
3	6	1	2	5	4
6	5	4	3	1	2
5	3	6	4	2	1
2	1	3	6	4	5

APPENDIX 2 (CONTINUED)

(2) 8 × 8 Latin squares
Suitable for:

Three samples and one standard each at two-dose levels, with a replication of eight for each test solution.

One sample and one standard each at two-dose levels, with a replication of sixteen for each test solution.

One sample and one standard each at two-dose levels, with two weighings of each and with a replication of eight for each test solution.

Design GF01

3	7	6	8	2	5	4	1
8	5	3	6	4	1	7	2
4	6	2	7	1	8	5	3
1	3	4	5	8	7	2	6
2	1	8	4	5	6	3	7
7	8	5	3	6	2	1	4
6	4	1	2	7	3	8	5
5	2	7	1	3	4	6	8

Design GF02

1	7	4	6	8	3	2	5
6	2	8	1	7	4	5	3
8	4	6	5	1	7	3	2
3	8	1	7	5	2	4	6
5	6	2	3	4	1	8	7
7	1	3	8	2	5	6	4
4	5	7	2	3	6	1	8
2	3	5	4	6	8	7	1

Design GF03

6	3	8	4	2	7	5	1
2	7	5	8	1	6	4	3
7	1	6	2	5	4	3	8
1	2	4	7	3	8	6	5
3	5	2	6	8	1	7	4
8	6	7	3	4	5	1	2
5	4	3	1	7	2	8	6
4	8	1	5	6	3	2	7

Design GF04

8	3	6	5	2	7	4	1
4	1	3	8	6	2	5	7
6	8	4	1	3	5	7	2
3	6	7	2	5	4	1	8
2	5	1	3	7	6	8	4
1	7	8	6	4	3	2	5
7	2	5	4	8	1	6	3
5	4	2	7	1	8	3	6

Design GF05

8	1	6	5	2	7	3	4
6	2	7	3	8	1	4	5
3	7	4	1	6	5	8	2
5	8	2	4	7	3	1	6
2	4	5	6	3	8	7	1
7	3	1	2	5	4	6	8
4	6	8	7	1	2	5	3
1	5	3	8	4	6	2	7

Design GF06

5	2	6	1	7	4	3	8
6	1	5	7	3	8	4	2
8	3	2	6	4	5	1	7
4	7	1	8	6	3	2	5
1	5	4	2	8	6	7	3
7	4	8	3	2	1	5	6
3	6	7	5	1	2	8	4
2	8	3	4	5	7	6	1

Design GF07

7	2	6	3	5	4	8	1
8	3	4	2	6	1	7	5
4	6	1	7	8	3	5	2
2	5	8	4	3	7	1	6
6	7	3	1	2	5	4	8
1	8	2	5	4	6	3	7
5	4	7	8	1	2	6	3
3	1	5	6	7	8	2	4

Design GF08

3	7	1	8	2	6	5	4
6	8	7	5	1	4	2	3
7	3	5	2	4	8	1	6
4	6	2	3	5	1	7	8
1	2	3	4	6	7	8	5
8	5	4	7	3	2	6	1
2	4	6	1	8	5	3	7
5	1	8	6	7	3	4	2

Design GF09

4	5	8	6	2	7	1	3
6	8	2	5	7	1	3	4
2	3	6	4	8	5	7	1
3	4	1	2	5	6	8	7
1	6	7	3	4	8	5	2
8	1	4	7	3	2	6	5
5	7	3	1	6	4	2	8
7	2	5	8	1	3	4	6

Design GF10

2	1	3	5	8	6	7	4
3	6	4	7	5	8	2	1
4	7	2	8	6	1	5	3
7	5	1	3	4	2	6	8
6	3	5	2	1	4	8	7
8	4	7	6	2	3	1	5
1	2	8	4	7	5	3	6
5	8	6	1	3	7	4	2

Design GF11

7	8	2	1	3	6	5	4
2	1	7	6	4	3	8	5
5	6	1	2	7	4	3	8
1	5	3	4	8	2	6	7
3	4	8	5	1	7	2	6
4	3	6	8	2	5	7	1
8	7	5	3	6	1	4	2
6	2	4	7	5	8	1	3

Design GF12

7	3	8	1	2	6	5	4
2	4	1	6	7	8	3	5
4	5	7	2	6	3	1	8
5	2	3	7	8	1	4	6
3	8	5	4	1	2	6	7
1	7	6	8	4	5	2	3
6	1	4	5	3	7	8	2
8	6	2	3	5	4	7	1

Design GF13

7	4	1	8	3	2	5	6
1	6	2	5	4	7	8	3
4	3	6	7	8	5	2	1
8	7	4	2	6	1	3	5
6	8	5	4	1	3	7	2
3	5	7	6	2	8	1	4
2	1	8	3	5	4	6	7
5	2	3	1	7	6	4	8

Design GF14

6	4	8	1	3	5	7	2
7	1	6	4	5	8	2	3
8	2	4	5	6	7	3	1
4	8	2	7	1	3	6	5
3	6	7	8	2	1	5	4
1	3	5	2	4	6	8	7
2	5	3	6	7	4	1	8
5	7	1	3	8	2	4	6

Design GF15

8	5	6	3	1	2	4	7
2	1	7	8	6	4	3	5
6	3	5	2	8	1	7	4
1	8	4	7	5	3	6	2
5	6	3	4	2	7	1	8
4	2	1	5	7	6	8	3
7	4	2	6	3	8	5	1
3	7	8	1	4	5	2	6

Design GF16

5	1	4	2	8	7	6	3
2	3	6	5	1	4	8	7
8	6	5	3	4	1	7	2
4	5	7	8	3	6	2	1
7	2	1	6	5	8	3	4
6	8	3	1	7	2	4	5
1	4	8	7	2	3	5	6
3	7	2	4	6	5	1	8

Design GF17

6	7	4	5	2	1	3	8
1	2	7	8	4	3	5	6
2	5	3	7	1	8	6	4
7	8	2	3	5	6	4	1
8	3	5	2	6	4	1	7
5	4	1	6	3	7	8	2
3	1	6	4	8	2	7	5
4	6	8	1	7	5	2	3

Design GF18

6	7	5	1	4	3	2	8
5	2	7	4	6	8	3	1
1	6	2	3	7	4	8	5
7	3	6	5	8	2	1	4
4	8	3	6	5	1	7	2
8	1	4	2	3	7	5	6
2	4	8	7	1	5	6	3
3	5	1	8	2	6	4	7

APPENDIX 2 (CONTINUED)

Design GF19

8	5	2	7	4	6	3	1
2	4	7	3	6	5	1	8
6	3	8	4	1	2	5	7
1	8	4	5	7	3	6	2
4	2	6	1	5	7	8	3
3	7	5	8	2	1	4	6
7	1	3	6	8	4	2	5
5	6	1	2	3	8	7	4

Design GF20

3	6	5	7	4	2	8	1
5	1	2	6	8	7	3	4
8	5	6	3	1	4	7	2
6	3	1	2	7	8	4	5
7	4	8	5	2	6	1	3
4	8	7	1	3	5	2	6
1	2	4	8	5	3	6	7
2	7	3	4	6	1	5	8

(3) 8 × 8 Quasi-Latin squares

Suitable for:

Seven samples and one standard each at two-dose levels, with a replication of four for each test solution.

Design GF01

14	10	7	11	1	6	3	16
8	4	13	5	15	12	9	2
15	13	2	4	12	9	8	5
12	2	5	13	9	8	15	4
6	16	11	7	3	14	1	10
9	5	4	2	8	15	12	13
1	7	16	10	6	3	14	11
3	11	10	16	14	1	6	7

Design GF02

13	15	12	2	8	3	5	10
10	12	13	5	3	2	8	15
7	1	6	16	14	9	11	4
1	7	4	14	16	11	9	6
4	6	7	11	9	16	14	1
15	13	10	8	2	5	3	12
12	10	15	3	5	8	2	13
6	4	1	9	11	14	16	7

Design GF03

7	9	13	4	16	12	1	6
16	4	12	13	1	9	6	7
6	12	4	9	7	13	16	1
3	5	15	8	14	2	11	10
10	2	8	5	3	15	14	11
1	13	9	12	6	4	7	16
14	8	2	15	11	5	10	3
11	15	5	2	10	8	3	14

Design GF04

16	9	13	6	7	4	12	1
6	1	7	16	13	12	4	9
13	4	6	7	16	1	9	12
7	12	16	13	6	9	1	4
2	5	11	10	3	8	14	15
11	8	10	3	2	15	5	14
10	15	3	2	11	14	8	5
3	14	2	11	10	5	15	8

Design GF05

8	16	3	13	10	1	6	11
4	12	7	5	2	9	14	15
6	10	13	3	16	11	8	1
15	5	12	2	7	4	9	14
14	2	5	7	12	15	4	9
1	3	10	16	13	6	11	8
9	7	2	12	5	14	15	4
11	13	16	10	3	8	1	6

Design GF06

2	15	11	14	5	7	4	10
10	11	15	4	7	5	14	2
12	9	1	8	3	13	6	16
16	1	9	6	13	3	8	12
7	4	14	15	2	10	11	5
3	8	6	1	16	12	9	13
13	6	8	9	12	16	1	3
5	14	4	11	10	2	15	7

Design GF07

8	11	1	4	13	16	10	5
9	6	14	7	12	3	15	2
2	15	7	6	9	12	14	3
5	10	4	11	8	13	1	16
13	4	10	1	16	5	11	8
12	7	15	14	3	2	6	9
3	14	6	15	2	9	7	12
16	1	11	10	5	8	4	13

Design GF08

9	2	12	15	5	14	3	8
6	7	13	4	16	1	10	11
4	13	7	6	10	11	16	1
11	16	10	1	13	6	7	4
8	5	3	14	12	9	2	15
15	12	2	9	3	8	5	14
14	3	5	8	2	15	12	9
1	10	16	11	7	4	13	6

Design GF09

3	5	16	13	9	8	12	2
10	4	1	6	14	15	7	11
8	12	5	2	16	13	9	3
2	16	9	8	12	3	5	13
15	7	4	11	1	6	14	10
6	14	7	10	4	11	1	15
11	1	14	15	7	10	4	6
13	9	12	3	5	2	16	8

Design GF10

6	15	10	1	7	12	3	14
15	12	3	6	10	1	14	7
4	13	2	9	11	8	5	16
12	1	14	15	3	6	7	10
9	4	11	8	16	13	2	5
13	8	5	4	2	9	16	11
8	9	16	13	5	4	11	2
1	6	7	12	14	15	10	3

Design GF11

16	10	2	5	13	7	12	3
5	3	13	10	12	2	7	16
14	8	4	11	15	9	6	1
11	1	15	8	6	4	9	14
1	11	9	14	4	6	15	8
3	5	7	16	2	12	13	10
10	16	12	3	7	13	2	5
8	14	6	1	9	15	4	11

Design GF12

14	15	8	9	6	1	11	4
9	14	15	8	1	4	6	11
12	3	2	5	10	7	13	16
3	2	5	12	13	10	16	7
5	12	3	2	7	16	10	13
8	9	14	15	4	11	1	6
2	5	12	3	16	13	7	10
15	8	9	14	11	6	4	1

Design GF13

16	9	13	8	11	2	6	3
9	2	6	13	8	3	11	16
7	14	4	1	10	5	15	12
12	7	1	10	15	14	4	5
3	16	8	11	6	9	13	2
14	5	15	4	1	12	10	7
5	12	10	15	4	7	1	14
2	3	11	6	13	16	8	9

Design GF14

11	14	16	7	5	4	2	9
1	8	6	13	15	10	12	3
13	12	10	1	3	6	8	15
4	5	11	16	2	7	9	14
16	9	7	4	14	11	5	2
10	15	1	6	12	13	3	8
6	3	13	10	8	1	15	12
7	2	4	11	9	16	14	5

Design GF15

3	14	7	2	12	15	5	10
7	10	3	12	2	5	15	14
2	5	12	7	3	14	10	15
16	11	6	13	9	8	4	1
13	4	9	6	16	11	1	8
12	15	2	3	7	10	14	5
9	8	13	16	6	1	11	4
6	1	16	9	13	4	8	11

Design GF16

16	6	3	13	10	7	1	12
5	9	8	2	11	14	4	15
1	3	12	6	7	16	10	13
10	12	13	3	16	1	7	6
4	8	15	9	14	5	11	2
7	13	6	12	1	10	16	3
11	15	2	8	5	4	14	9
14	2	9	15	4	11	5	8

APPENDIX 2 (CONTINUED)

Design GF17

8	14	5	16	1	11	3	10
9	7	2	13	12	4	6	15
11	1	8	5	10	16	14	3
13	15	4	9	6	2	12	7
16	10	11	8	3	5	1	14
5	3	16	11	14	8	10	1
4	12	9	2	15	13	7	6
2	6	13	4	7	9	15	12

Design GF18

3	2	11	7	13	6	16	10
5	14	15	9	1	4	12	8
1	4	9	15	5	14	8	12
13	6	7	11	3	2	10	16
12	9	14	4	8	15	1	5
8	15	4	14	12	9	5	1
16	7	2	6	10	11	13	3
10	11	6	2	16	7	3	13

Design GF19

7	16	1	14	10	5	4	11
10	7	4	1	11	14	5	16
6	3	8	15	13	12	9	2
13	6	9	8	2	15	12	3
3	2	15	12	6	9	8	13
16	11	14	5	7	4	1	10
2	13	12	9	3	8	15	6
11	10	5	4	16	1	14	7

Design GF20

10	6	13	1	3	15	12	8
5	7	4	16	2	14	9	11
15	13	12	6	10	8	1	3
2	16	7	9	11	5	4	14
11	9	16	4	14	2	7	5
3	1	6	12	8	10	13	15
8	12	1	13	15	3	6	10
14	4	9	7	5	11	16	2

APPENDIX 3
Tests for outliers by two *USP* criteria. These were introduced in "Detection of Outliers" in Chapter 8.

Part A

Test for outliers by *USP* criterion 1

N	3	4	5	6	7						
G_1	0.976	0.846	0.729	0.644	0.586						
N	8	9	10	11	12	13					
G_2	0.780	0.725	0.678	0.638	0.605	0.578					
N	14	15	16	17	18	19	20	21	22	23	24
G_3	0.602	0.579	0.559	0.542	0.527	0.514	0.502	0.491	0.481	0.472	0.464

This is Table 1 of *USP 24* <111>, Design and Analysis of Biological Assays. In samples from a normal population, gaps equal to or larger than the following values of G_1, G_2, and G_3 occur with a probability $P = 0.02$ where outliers can occur only at one end, or with $P = 0.04$ where they may occur at eiher end.

APPENDIX 3 (CONTINUED)

Part B
Test for outliers by USP criterion 2.

Part 1, critical R^* for ranges each from f observations.

Number of Observations, $f \rightarrow$		4	5	6	7	8	9	10
Number of ranges, k	4	0.479	0.446	0.425	0.410	0.398	0.389	0.382
	5	0.398	0.369	0.351	0.338	0.328	0.320	0.314
	6	0.342	0.316	0.300	0.288	0.280	0.273	0.269
	7	0.300	0.278	0.263	0.523	0.245	0.239	0.234
	8	0.267	0.248	0.234	0.225	0.218	0.213	0.208

Part 2, critical $(k + 2)$ R^* ranges each from f observations.

Number of Observations, $f \rightarrow$		2	3	4	5	6	7	8
Number of ranges, K	10	4.06	3.04	2.65	2.44	2.30	2.21	2.14
	12	4.06	3.03	2.63	2.42	2.29	2.20	2.13
	15	4.06	3.02	2.62	2.41	2.28	2.18	2.12
	20	4.13	3.03	2.62	2.41	2.28	2.18	2.11
	50	4.26	3.11	2.67	2.44	2.29	2.19	2.11

For both parts 1 and 2, (a) compute the range from the smallest to the largest observations in each treatment group; (b) compute the ratio R^* of the largest range to the sum of all ranges.

For part 1, if R^* equals or exceeds the number corresponding to f and k, the largest range is suspect and may contain an outlier ($P = 0.05$).

For part 2, if $(k + 2)R^*$ equals or exceeds the number coresponding to f and k, the largest range is suspect and may contain an outlier ($P = 0.05$).

When a range is suspect, inspection will usually identify the apparent outlier.

This is a part of Table 2 of *USP 24* <111>, Design and Analysis of Biological Assays.

APPENDIX 4
Variance Ratio Tables — The *F* Test

Variance ratios were introduced in Chapter 9.

0.1% Probability

n_2 \ n_1	1	2	3	4	5
1	405,284.0	500,000.0	540,379.0	562,500.0	576,405.0
2	998.5	999.0	999.2	999.2	999.3
3	167.0	148.5	141.1	137.1	134.6
4	74.1	61.3	56.2	53.4	51.7
5	47.2	37.1	33.2	31.1	28.8
10	21.0	14.9	12.6	11.3	10.5
15	16.6	11.3	9.3	8.3	7.6
20	14.8	10.0	8.1	7.1	6.5
30	13.3	8.8	7.1	6.1	5.5

1% Probability

n_2 \ n_1	1	2	3	4	5
	4,052.0	4,999.0	5,403.3	5,625.0	5,764.0
	98.5	99.0	99.2	99.3	99.3
3	34.1	30.8	29.5	28.7	28.2
4	21.2	18.0	16.7	16.0	15.5
5	16.3	13.3	12.1	11.4	11.0
10	10.0	7.6	6.6	6.0	5.6
15	8.7	6.4	5.4	4.9	4.6
20	8.1	5.9	4.9	4.4	4.1
30	7.6	5.4	4.5	4.0	3.7

5% Probability

n_2 \ n_1	1	2	3	4	5
1	161.4	199.5	215.7	224.6	230.2
2	18.5	19.0	19.2	19.3	19.3
3	10.1	9.6	9.3	9.1	9.0
4	7.7	6.9	6.6	6.4	6.3
5	6.6	5.8	5.4	5.2	5.0
10	5.0	4.1	3.7	3.5	3.3
15	4.5	3.7	3.3	3.1	2.9
20	4.4	3.5	3.1	2.9	2.7
30	4.2	3.3	2.9	2.7	2.5

APPENDIX 4 (CONTINUED)

10% Probability

n_1	1	2	3	4	5
n_2					
1	39.9	49.5	53.6	55.8	57.2
2	8.5	9.0	9.2	9.2	9.3
3	5.5	5.5	5.4	5.3	5.3
4	4.5	4.3	4.2	4.1	4.0
5	4.1	3.8	3.6	3.5	3.5
10	3.3	2.9	2.7	2.6	2.5
15	3.1	2.7	2.5	2.4	2.3
20	3.0	2.6	2.4	2.3	2.2
30	2.9	2.5	2.3	2.1	2.1

20% Probability

n_1	1	2	3	4	5
n_2					
1	9.47	12.00	13.06	13.64	14.01
2	3.56	4.00	4.16	4.24	4.28
3	2.68	2.89	2.94	2.96	2.97
4	2.35	2.47	2.48	2.48	2.48
5	2.18	2.26	2.25	2.24	2.23
10	1.88	1.90	1.86	1.83	1.80
15	1.80	1.79	1.75	1.71	1.68
20	1.76	1.75	1.70	1.65	1.62
30	1.72	1.70	1.64	1.60	1.57

Source: © R.A. Fisher and F.W. Yeats. 1963. Reprinted by permission of Pearson Education Ltd.

APPENDIX 5

Parts of Table 9 of the *USP 24* <111>, Design
and Analysis of Biological Assays. This table
was introduced in Chapter 9, "Example 9.1,
Continued — The Calculation According to
USP Guidance."
Values of F_i for different degrees of freedom, n
that will be exceeded with a probability $P =$
0.05 (or 0.95 for confidence intervals).

n	F_1	F_2	F_3
1	161.450	—	—
2	18.510	19.00	19.16
3	10.128	9.55	9.28
4	7.709	6.94	6.59
5	6.609	5.79	5.41
6	5.987	5.14	4.76
20	4.351	3.49	3.10
21	4.325	3.47	3.07
22	4.301	3.44	3.05
23	4.279	3.42	3.03
24	4.260	3.40	3.01
25	4.242	3.38	2.99
26	4.225	3.37	2.98
27	4.210	3.35	2.96
28	4.196	3.34	2.95
29	4.183	3.33	2.98

APPENDIX 6
The *t* Distribution

Student's *t* was introduced in Chapter 9, "A Three-Dose-Level Assay for One Unknown." The table gives the value *t* corresponding with various values of the probability of a random value falling outside the limit $\pm t$.

The values tabulated here refer to $P = 0.05$ only.

n	t	n	t
1	12.706	11	2.201
2	4.303	12	2.179
3	3.182	15	2.131
4	2.776	20	2.086
5	2.571	25	2.060
6	2.447	30	2.042
7	2.365	40	2.021
8	2.306	60	2.000
9	2.262	120	1.980
10	2.228	infinity	1.960

Source: © R.A. Fisher and F.W. Yeats. 1963. Reprinted by permission of Pearson Education Ltd.

APPENDIX 7

PATTERNS FOR THE DISTRIBUTION OF TEST SOLUTIONS ON SMALL PLATES

A petri dish normally accommodates six test solutions. It is convenient to number the test solutions 1 to 6 and then distribute them on the assay plate in accordance with a template. Four suitable templates are shown below.

Schemes for the numbering of solutions are shown for 2 + 2 and 3 + 3 assays:

2 + 2 assays

	Unknown 1	Unknown 2	Standard
Low dose	1	3	5
High dose	2	4	6

3 + 3 assays

	Unknown	Standard
Low dose	1	4
Mid dose	2	5
High dose	3	6

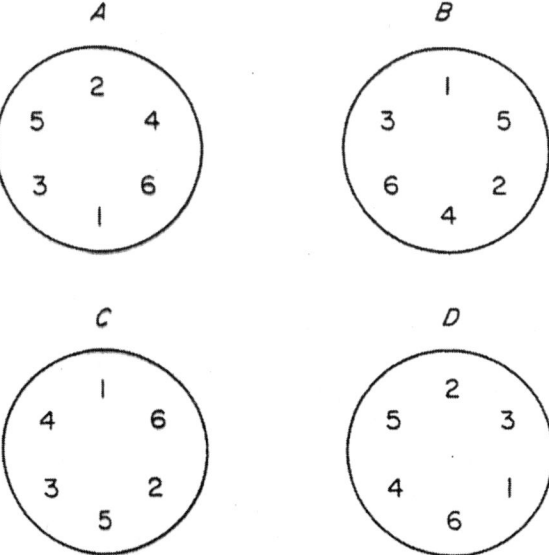

FIGURE A Patterns suitable for use with 2 + 2 assays or 3 + 3 assays.

Each plate accomodates only two test solutions that are applied in triplicate.

Position R is for the reference solution, which is the mid dose of the five standard doses. The same test solution is applied in position R for every plate in the assay.

Position S is for a single dose of the unknown or for any one of dose levels 1, 2, 4, or 5 of the reference standard. It is customary to use three plates for each unknown and three plates for each of dose levels 1, 2, 4, and 5 of the reference standard.

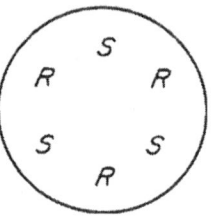

FIGURE B A pattern for use in 5 + 1 assays.

APPENDIX 8

Evidence that a quadratic component of curvature in the log dose-response line in a symmetrical, three-dose-level parallel-line assay does not invalidate the assay.

Assumptions:

1. The standard procedure of the *International* (*IP*) and other pharmacopoeias for computation of potency estimate in the agar-plate diffusion assay for antibiotics is based on a log dose-zone size line represented by

$$y = a + bx \qquad (A8.1)$$

where x is log dose (or coded log dose), y is the mean of observed zone sizes (usually in terms of diameter) corresponding to that dose, a is a constant and b is the slope as defined in assumption 3, which follows.

2. For the purpose of this argument, it is supposed that, rather than equation (A8.1), the line is better represented by:

$$y = \alpha + \beta x + \gamma x^2 \qquad (A8.2)$$

where x and y have the same meanings as before; α, β, and γ are constants.

3. The symbols used in the computation procedure are those of the *International Pharmacopoeia* thus:

S_1, S_2, and S_3 are mean responses to low, medium, and high dose of standard, respectively.
T_1, T_2, and T_3 are mean responses to low, medium, and high dose of unknown, respectively.

> E = difference in mean response between adjacent dose intervals.
> F = difference in mean response between standard and unknown at the same dose level.
> I = logarithm (base 10) of ratio of doses between adjacent levels.
> b = the calculated (extrapolated) increase in mean zone size corresponding to a tenfold increase in dose.
> M = $\log_{10}$ of the estimated potency of the unknown (i.e., ratio of estimated potency of unknown to that of the standard).

In addition to the foregoing symbols, we introduce:

Q = logarithm of the true ratio of potency of unknown to that of the standard.

PRINCIPLE OF THIS ARGUMENT

Using Equation (A8.2), theoretical responses (lying on the curved response line) are calculated to the three assumed doses of each standard and unknown preparation. The six values are then substituted in the standard expressions for computation of potency estimate. The calculated logarithm of potency estimate, M, is then compared with the known logarithm of potency, Q.

PROOF

Setting logarithm standard mid dose as zero, the complete set of values of x and y are:

for standard $\quad x \quad\quad\quad y$

$$-I \quad\quad S_1 = \alpha - \beta I + \gamma I^2$$
$$0 \quad\quad S_2 = \alpha$$
$$+I \quad\quad S_3 = \alpha + \beta I + \gamma I^2$$

for unknown $\quad x \quad\quad\quad y$

$$Q - I \quad\quad T_1 = \alpha - \beta(Q - I) + \gamma(Q^2 - 2QI + I^2)$$
$$Q \quad\quad T_2 = \alpha + \beta Q + \gamma Q^2$$
$$Q + I \quad\quad T_3 = \alpha + \beta(Q + I) + \gamma(Q^2 + 2QI + I^2)$$

Working out the values for these theoretical responses and then substituting them in the standard expressions for E and F, it is found that:

$$E = I(\beta + \gamma Q) \text{ and } F = Q(\beta + \gamma Q)$$
$$\text{then } b = E/I = I(\beta + \gamma Q)/I = \beta + \gamma Q$$
$$\text{and } M = F/b = Q(\beta + \gamma Q)/(\beta + \gamma Q) = Q$$

Thus, the estimated logarithm of potency ratio, M, is identical with the known logarithm of potency ratio, Q, even though logarithm of dose and response were related by an expression having a quadratic component of curvature.

Note: Appendix 8 is taken from a publication by the author and is reproduced verbatim with the kind permission of the World Health Organization.

APPENDIX 9
The Distribution of X^2

The use of this function was introduced in Chapter 12, *"International Pharmacopoeia Guidance."*

P	0.20	0.10	0.05	0.01
n				
1	1.64	2.71	3.84	6.64
2	3.22	4.61	5.99	9.21
3	4.64	6.25	7.82	11.35
4	5.99	7.78	9.94	13.28
5	7.29	9.24	11.07	15.09
6	8.56	10.65	12.59	16.81
7	9.8	12.02	14.07	18.48
8	11.03	13.36	15.51	20.09

Index